PROJECT MANAGEMENT 2.0

LEVERAGING TOOLS, DISTRIBUTED COLLABORATION, AND METRICS FOR PROJECT SUCCESS

Harold Kerzner

INTERNATIONAL
Institute for Learning, Inc.

WILEY

Cover images: Gold Guy Workgroup © Fotolia/Scott Maxwell; Brushed Metal Plate © iStock.com/Zeffss1
Cover design: C. Wallace

This book is printed on acid-free paper.

Published by John Wiley & Sons, Inc., Hoboken, New Jersey

Published simultaneously in Canada

For general information about our other products and services, please contact our Customer Care Department within the United States at (800) 762-2974, outside the United States at (317) 572-3993 or fax (317) 572-4002.

Wiley publishes in a variety of print and electronic formats and by print-on-demand. Some material included with standard print versions of this book may not be included in e-books or in print-on-demand. If this book refers to media such as a CD or DVD that is not included in the version you purchased, you may download this material at http://booksupport.wiley.com. For more information about Wiley products, visit www.wiley.com.

Library of Congress Cataloging-in-Publication Data:

ISBN 978-1-118-99125-1 (pbk); ISBN 978-1-118-99128-2 (epdf); ISBN 978-1-119-00028-0 (epub); ISBN 978-1-119-02004-2 (WOL)

Printed in the United States of America

10 9 8 7 6 5 4 3 2

DEDICATION

To my wife, Jo Ellyn, for her encouragement to continue writing project management books, her patience with my travel schedule, and, most important, her everlasting love.

CONTENTS

CHAPTER 11
NEED FOR PROJECT MANAGEMENT   283

CHAPTER 12
USING THE PMO TO SPEARHEAD PM 2.0   295

The world is in a state of constant flux. We see the changes happening every day on television, in the newspapers, and on the Internet. But how many people actually recognize the changes that are taking place in project management? For those individuals who live, eat, and breathe project management, they often are not able to recognize the changes even though they are part of the change.

When project management (PM) first was introduced, senior management was somewhat apprehensive about accepting this new technique. Even though project management had existed as long as mankind and used the principles of PM 1.0, executives viewed project management as a special form of management that was more temporary than permanent. Clients were forcing corporations to use project management for the life of the projects so that the clients would have single-person contact concerning the performance of the projects. Reluctantly, senior management for the contractors accepted the challenge, but there was some fear that project managers might usurp the authority of the position and begin making decisions that were reserved for C-level personnel.

In order to maintain some degree of control, senior management created the position of project sponsor. Whatever decisions project managers were allowed to make would be under the watchful eyes of the project sponsors. In many instances, the project managers came from the engineering ranks of the company and were allowed to make mainly technical decisions. Virtually all business-related decisions were made by the executive sponsor. Furthermore, many sponsors maintained the single-person executive–client contact link with the customer rather than allowing the project manager to have free access to the customers.

Today, all of that has changed thanks to PM 2.0. We now live in a world of PM 2.0. The survival of many companies is now based upon how well they meet the challenges in the marketplace. Many of the challenges include greater business risks, having to accept more complex projects, and working closer with a multitude of stakeholders. Project management processes must be in place to meet these challenges.

For the past three decades, companies have embarked upon continuous improvement efforts in all areas of project management. Although some changes were large, most of the changes were small, even just cosmetic, and usually involved the forms, guidelines, templates, and checklists we use for project management execution. The projects in most companies were regarded as operational rather than strategic projects. Today, much of this has changed.

PM 2.0 has now spread to the seniormost levels of management and even to the corporate boardroom. Projects are now being aligned with strategic planning objectives. Project management is being used to execute strategic projects rather than just operational projects. Project managers are expected to make both project-related and business-related decisions whereas in the past it was only a project-based decision. Today's project managers are viewed as managing part of a business instead of merely a project.

The marriage of project management with business analyst activities has elevated project management to the corporate level. Project management is now seen as a strategic competency and necessary for the survival of the firm. Project managers no longer report to just a project sponsor. Instead, they report to a senior governance committee, an oversight committee, or the seniormost levels of management. Project sponsorship is now committee governance rather than oversight by a single individual. This is because of the risks and complexities of today's projects.

This book is designed to be forward looking. Project management has undergone numerous changes in the last few years and many of the changes appear in this book. Change is inevitable. It will happen, and it will continue to happen. Whether you accept the changes now or in the future is up to you. But change is a way of life.

Harold Kerzner
The International Institute for Learning
2014

ACKNOWLEDGMENTS

The vision of the future is held in the hands of many rather than just a few. The idea for a book on PM 2.0 came from Seppo Halminen, who continuously challenges me with new ideas surrounding the future of project management. The author is indebted to Greg Balestrero, John Winter, and Carl Belack for their invaluable contributions to this book. The author is also indebted to all of the employees of the International Institute for Learning who have provided me with 25 years of support for my books.

FOREWORD

Knowledge is an unending adventure at the edge of uncertainty.
Jacob Bronowski

Story telling is one of the greatest ways of imparting knowledge and sharing traditions and culture. So let me start by sharing a personal story about my youthful obsession with reading. Science fiction writing has intrigued me since I was a very young boy. I read every sci-fi writer I could get my hands on and I could understand. To me, these writers allowed me to look through their eyes into the future. These writers were my first exposure to "futurists," showing me a future that might, and could, someday be played out. It made me dream big.

They were great dreamers, like Ray Bradbury, Isaac Asimov, and Arthur C. Clarke. Even before the industrial age got into full swing, Jules Verne, in 1865, cast a wonderful story about our first journey to the moon, a dream that would come true 100 years later. They were a small group of writers who were instrumental in my first big dream about the future: *I wanted to become an astronaut.* From when I was 8 years old I had no other dream for a future except that I would ultimately be in a space suit, walking on another planet. Sadly, destiny intervened when, at age 20, a U.S. Air Force optometrist informed me that I would wear corrective lenses for the rest of my life! Well, pilots at that time had to have perfect vision, so my dream of becoming an astronaut was gone. And, the rest is, as they say, history.

To me, this seemed like a shock. For the next 2 years, I wandered through the Georgia Institute of Technology taking a variety of courses in literature, philosophy, and science, before landing on a major in industrial engineering. As I look back, I learned a lesson that it is like many things in life—it is about balancing our vision, personal expectations, and looming reality or we will be knocked back on our heels by the outcome. I should have known the outcome would be unlikely. In fact, if you saw an early picture of my family together, including parents, uncles, aunts, and some cousins, you would make an interesting observation. It was easy to see that everyone was wearing glasses, except for an 8-year-old, named Greg. Hmm, there was a "trend" going on that I ignored because of my deep-rooted, passionate belief about being an astronaut. Someone might even say I was ignoring the "inside" facts.

However, there was another "outside" trend that was happening, especially as I sat in college wondering what would happen to my interest in being in space and mapping it against a future career. It was a period just prior to the U.S. Apollo mission landing two astronauts on the moon. The war in Viet Nam was raging, the United States was embroiled in its own cultural revolution, and the U.S. congress was feverishly debating the funding for both swords and plowshares. Projections about the following 10 years were beginning to predict the end of the Space Race, an end to the war, and a looming recession. As a 20-year- old, I didn't see those "outside" trends resulting in the demise of the aerospace industry of the time. Within 5 years, employment in the aerospace industry collapsed from nearly 1 million professionals to fewer than 200,000. The war ended and the need for new military pilots collapsed. And the recession dried up opportunities in many career fields. Apparently, the "crushing blow" of losing the opportunity of becoming an astronaut may, in fact, have been a good result. I would have had a degree in aerospace engineering and no job potential in the field. In this case, the outside forces showed me that wearing glasses was not such a bad deal after all!

WHY THIS STORY MAKES SENSE

So, why would I share this story in a book on PM 2.0 by an esteemed and successful writer like Dr. Kerzner? Simply I wanted to demonstrate that an awareness of the inside, local, and outside forces under which we will work, whether 6 months or 5 years ahead, will be one of the most important contributors to our success as leaders. As recently as 2007, the world was flying high with the highest prosperity in our history. However, it was evident that a massive market adjustment was coming in the fall. No one knew how big, but by the end of 2007, the entire global economy receded greater and more abruptly than anything since the great global depression of 1929. More projects were canceled abruptly than any time in the history of tracking project execution. And, you guessed it, more project managers were out of work than ever before since the end of World War II.

Unfortunately, as project professionals, we recoil at the thought of uncertainty. Good project managers are driven to find the nodes of uncertainty and create a mitigation plan to continue to operate through these uncertain times. And, yet, the term "uncertain" is, by definition, not sharply defined and does not offer one clear answer for a strategy. My example above demonstrates that there was a certainty about my dream that I was unwilling or incapable of understanding as a child. Even as an adult, I didn't see the possible collapse of the aerospace industry. As project professionals, we don't have the luxury or the time to ignore reality and keep our heads down to the triple constraints of any project. As leaders, we must breed a culture of awareness of the world around us, under which we must operate.

THROUGH THE LOOKING GLASS AT A CHAOTIC FUTURE

We have all heard too many times in our recent past that we are in chaotic times, crazy times, and complex times. And, yet, we haven't faced the most chaotic in human history. Somehow, we have always managed to move our society ahead, with advances in technology to make our lives more manageable; new innovations in medical care to eradicate horrible diseases and provide a firewall to pandemics; and persistent and effective efforts to move more and more people out of poverty. However, the undercurrents in the twenty-first century are pointing to a very chaotic time. Let's take a closer look at the pending world in the mid-twenty-first century and see if our own expectations match the real world. More importantly, are we missing something that would affect the way we work and companies for which we work?

The first issue, and probably the most significant, is *population growth* and its impact on a global society. The global population at the end of 2014 is estimated to be 7.125 billion. It is a very dynamic number, with the global society adding a net increase of one person to the population every 13 seconds. At the current rate, it means that by the year 2050 we will exceed 9 billion in population. Let's put that in perspective. The first billion in population was reached in the first decade of the 1800s. That was the accumulated growth of all of human history. The second billion came 130 years later in the early 1930s. That is when things really changed. Since then, the average increase has been another billion every 14 years! This growth is making it much more difficult to provide food, water, clothes, housing, and amenities for the masses.

Another issue is that there is a *rush to the cities*. Earlier in the last decade, the world's demographic shifted in a subtle but dramatic way. For the first time in human history, slightly more than 50% of our population resided in cities, and by 2050, it is estimated by the United Nations (UN) that 70% of the world's population will live in cities. To put that into perspective, in 2014, approximately 3.5 billion are living in cities. In 2050, that number will be nearly 6.5 billion. These people will be living in cities that are not around in 2014. China alone wants to move more than 275 million people into cities from the agricultural regions by 2020.

Related to the growth in population and cities is the *regional source for this growth*. The growth is focused on Africa, Asia, China, and South America. These regions of the world are in turbulent development phases, while the developed regions are slowing in growth, and future demographics are balancing out. Regardless of what is done to slow the growth in developed nations, it will never offset the growth in the developing regions of the world.

And let's not forget *climate change*. This is an outlier that is finally being debated on a daily basis. Nearly every major company in the world sees climate change as the most serious environmental issue we face. Due to greenhouse gases discharged into the atmosphere through human activities such as the operation of transportation vehicles, electric power generation, manufacturing facilities, water treatment facilities, and more, the impact on climate is significant. The warming trend is causing an increase in the mean tidal height of the seas, estimated to be 39 inches higher on average, on the coastal regions of the United States alone. Bad news? Estimated financial damage in today's dollars is around $1.5 trillion. That same condition is causing shifts in climates affecting crop yields across the globe, whether it is for soft drink sweeteners, hops and barley for beer, vegetables and fruits, cocoa for chocolate, or crop sources for ethanol production.

The *availability of fresh water* is another critical condition that we face. In the developed world, we are spoiled by the fact we can open a water tap and get a glass of water with full confidence that we will get water that is safe and clean to consume. Perhaps it tastes odd because of some chemical such as chlorine, but by and large, we can drink it, cook with it, bathe with it, and not worry about polluted water or water-borne diseases. However, that is not the case for nearly 50% of the world's population. The UN has reported that around 1.2 billion people live in areas of physical scarcity of water and another 500 million are approaching water scarcity. They also report that another 1.6 billion people face economic water shortage where the country lacks the wherewithal to take water, clean it, and deliver it to the populace. However, the real issue is that the demand for water has been growing at twice the rate of growth of the population! In reality, we have enough water today for about 7 billion people, not the 9 billion we will see in 2050.

And, of course, our favorite subject to argue about: *energy*. Next to climate change, energy production, particularly as it is associated with carbon-based fuels, is the most hotly debated subject in the world. There is no doubt that we have reached a turning point in the easily accessible carbon-based fuels, especially crude oil. Shell Oil, in the latest publication of its future scenarios, describes this era as the retirement age of carbon-based fuels. In fact, the Shell scenarios predict that the last internal combustion engines running on carbon-based fuels will be made in 2070. Throughout the world, there is turmoil to determine the most effective policy mix of renewable and nonrenewable energy sources, whether it is for transportation, for electricity production, or for other scenarios.

IS IT HALF EMPTY OR HALF FULL OR JUST PLAIN COMPLICATED?

These five trends are all significant in and of themselves. So, can't we just, as Steve Jobs said at Apple, "innovate our way out of the situations?" Well, before one can answer the question, you need to know that each of these five megatrends are interrelated, creating a highly complex set of problems in the future. It is impossible to give full treatment, but let me "sing a few bars of this tune" and let you be the judge:

- Drinking and cooking represents only a small portion of the uses of water. In fact, the three biggest consumers of water are agriculture (irrigation), energy (cooling), and manufacturing (cooling, cleaning, product use, etc.). Of the 2.5% of the water on the planet that is fresh, only 20% of it is readily accessible and available to meet these multiple needs. Growing populations on the planet create an ever-increasing demand for food, energy, and "things"—all of which need water as a raw material. So, essentially our human population is in a tug of war for water due to the very demands that we create.
- There is enough water on the planet now for 7 billion people, but we are going to grow to 9 billion people in about 30 years. Keep in mind that the amount of water on the planet is fixed—we can't produce more water, ship any in from the moon or the planets, or "grow" it. The short story is that if we proceed with a "business as usual" consumption rater, water scarcity will only spread across the globe.
- Today, climate change is affected by greenhouse gas emissions from virtually every point on the compass rose. However, with projected population growth and a massive change in city growth, the demand for wood products, steel, and other construction products will skyrocket, especially in the developing regions of the world. The U.N.'s Food and Agriculture Organization (FAO) estimates that over 13 million hectares (30 million acres) of forests are destroyed by human activity, whether for raw materials or for clearing land for housing and cities. As more trees are removed from the carbon sequestration process, more carbon emissions stay in the atmosphere, and climate change becomes more aggressive and chaotic.
- With a more aggressive and chaotic climate change, flooding and droughts will negatively affect crop yields, at a time when the demand for food will only go up.

Efforts to improve crop yields open the door to ethical debates on genetically modified crops, improper agricultural labor practices in developing nations, and a host of other issues.

- And the urgency of "want" is creating unrest throughout the world. In the developed world, we see the freedom from want as a right, an inalienable right. But for most of the world, it is a dream, out of reach. Fresh water, sufficient food to keep from being hungry, shelter, sanitation facilities, a meaningful job, and more still remain out of the grasp of the majority of our population. Just 15 minutes of watching global news broadcasts will awaken one's sensitivity to this global unrest.

Couple these challenges with the demands for good and services of a massively growing global population and the issue becomes even more complicated, reaching somewhat of a crisis stage. Consider industries related to information and communications technology (ICT) as an example. The demand for mobile communication and computing has gone up exponentially. In the 2012 report of the International Telecommunications Union (ITU), the number of cell phone users in the world was listed at 6 billion! In fact, 105 countries had more cell phone users than inhabitants. More importantly, there were more people in the world with cell phones than with access to fresh drinking water and sanitation facilities!

This is increasing, particularly in the area of smart phones and tablets. The annual growth rate for smart phones in 2013 was 44%, growing to nearly 1.5 billion users worldwide. With a growing population, especially in the cities, it is expected that this number will continue to grow dramatically. With a life cycle of 6–12 months, the number of cell and smart phones that must be recycled is going up exponentially as well.

A hidden issue behind this phenomenal growth in the ICT industry is the demand for unique and critical raw materials. The rare earth elements (REEs) and a variety of other metals (e.g., copper, aluminum, cobalt) are in such demand today that the prices continue to climb while the demand for cheaper and cheaper products goes up as well. Ninety-five percent of the worlds' supply of REEs originate in the People's Republic of China (PRC). In fact, Deng Xiaoping, the architect moving China toward a market economy, once observed that these minerals will be to China what oil is to Saudi Arabia. Restrictions to the distribution of the REEs and increasing prices have made that prediction a reality. This has led to the production of some REEs (e.g., tantalum) in other countries, such as the Kingdom of Saudi Arabia (KSA) and the Democratic Republic of the Congo (DRC), but not without its own problems. The sale of tantalum from the DRC has been funding brutal civil wars and gradually being considered a "blood mineral."

SO WHAT DOES ALL THIS MEAN TO YOU?

So, you are probably reading this and asking how and why ethics, politics, import and export restrictions, and responsibility for supply chains can affect your day-to-day activities as a project professional. Can you sense the complexity and interrelationship of all of these risks? Trust me when I say I could share similar challenges in virtually every industry. But, more importantly for you, have you begun to wonder what impact all of this has on your company and your career? Well, if the last question is the one that is swimming in your head, then you are asking exactly the right question to prepare you to become a sound and pragmatic project management professional in the future.

Regardless of the company or industry you are associated with, the next 40 years will provide fertile ground for strategic changes that will affect every aspect of the operations and delivery of companies. Or, to put it differently, if a company doesn't change dramatically to address these global challenges, then the company will soon be closing its doors forever—whether it is 5 years, 10 years, or 15 years out, no change means a death toll to the company.

And yet there is hope. Companies around the world are heeding these environmental and societal risks. More than 10,000 companies, nongovernmental organizations (NGOs), universities, and more have signed on to the UN Global Compact to address 8 Millennium Development Goals which attack our worldwide problems with climate change, poverty, empowerment of women, energy efficiency, global health. and more. More than 45,000 companies have joined the Global Reporting Initiative (GRI), cosponsored by the UN, to aid companies in setting and reporting progress on specific goals to address the challenges I mentioned earlier. In a review of dozens of the 10-K reports which publically traded companies are required to file in the United States to the Securities and Exchange Commission, we found a remarkable list of future risks which parallel what I have shared here. It is the same throughout the world. Yes, there is hope, and companies are changing.

But you, as a citizen, employee, and project professional need to step up and take action to help your organization, community, and the world address these challenges. Will you be ready for that challenge? When thinking about the conditions under which you will be working in the future, you need to embrace that every company and associated projects will have the following constraints:

- A holistic, zero-waste approach will rule our work. Waste will not have a place in a future of constrained resources. In the past, we placed a narrow focus on waste to the budget and to schedule or time, constrained by the scope of the project—never go over budget, never waste time. However, in the future, we will address every resource in the process, whether it is water, raw materials, or talent. New metrics embracing these and other constrained resources will rule the day.
- Water will be considered one of the most precious resources and will innovate ways to become water neutral. Innovation will rule the day and products, processes, and operations will be redesigned to keep the limited amount of fresh, potable water at the same level as it was before it entered your accountability stream.
- All companies must and will embrace their "carbon footprint" and not only find ways to reduce the footprint in all of its activities but also contribute to the ways that carbon can be reabsorbed. It is a complex science but it will affect the everyday work of all professionals.
- Expect increased exposure to a host of stakeholders that currently are not in the mix. The term "customer" will expand greatly in meaning and focus as companies try to address the needs of a broader community and society at large. Thanks to the visibility and accountability that social media brings, you can't hide or "spin" mistakes. If the future, you will find yourself on a stage with an

audience of thousands of individuals from the local communities in which you operate.

- Companies will take full responsibility and be accountable for the entire value chain. Every supplier that adds value to the deliverables of the company will be the company's responsibility, whether they are a contractor or a division-owned company.

- Ethics and shared values must be a part of your everyday responsibilities as a professional. Wherever I speak on the subject of ethics, I ask the professionals in the audience about the discussions they have with colleagues, whether it is when they are briefing new employees or beginning a new project. In particular, I ask how many of them allocate time talking about the ethical implications of their work, values that will be embraced in the execution of the work, and what to do about an ethical "moment of clarity." The average answer is less than 5%. In the future, the answer should be 100%, or there will be moments of clarity that will leave you flat footed and open to failure.

- Risk management must be considered an enterprise competency. This means that the risk register associated with the execution of a project must contain the "strategic risks" to the company. If the availability of fresh water is a strategic risk to the company, then it must somehow fall into the purview of project, clearly and knowingly. If it does not, then your project is at risk in a way that will "blind side" the success of the project.

- Successful companies in the future will find a way to build risk "radar" into the everyday responsibilities of your work. Expect to be asked to be "aware" of what is going on outside the window of your building or the monitor of your computer. Risk "radar" will keep the company from the shock and awe than can be dished up from a chaotic future.

Well, enough about the future. The real question is what skills and abilities do I need as a successful project professional to work in the future. Well, that is the purpose of this book. PM 2.0 is the best guide you can have in your hands to cope with the future. There are no more experienced or knowledgeable professionals than Harold Kerzner to help you get ready for tomorrow. He is one the most important contributors of our time to the project management body of knowledge. As you will see throughout the book, Dr. Kerzner brings PM 2.0 to life in a variety of applications and through the myriad of knowledge areas and processes deployed in the execution of a project. However, most of all, he is lighting a pathway to increase your value to the company and to its stakeholders. He provides navigational aids whether you are working with virtual teams, in R&D, or leading one of the most significant strategic projects in your company. In short, what you have in your hands is an operational guide to the most complex human enterprise you will encounter: the project enterprise. My only advice is to read it, take it seriously, and keep it within arms reach at all times. As they say in the American Express commercial, "never leave home without it." It is your manual for success.

Greg Balestrero is the Strategic Advisor on Leadership, Sustainability & Corporate Consciousness for the International Institute for Learning. He was formerly the CEO of the Project Management Institute.

CHAPTER 1
PROJECT MANAGEMENT 2.0

1.0 INTRODUCTION: CHANGING TIMES

In today's business environment, we have a new generation of workers that has grown up in a Web 2.0 world of Web-based project management tools allowing people on virtual or distributed teams to work together much more closely than in the past. Advances in computer technology and information flow have shown that the way we traditionally managed projects, PM 1.0, is a hindrance and ineffective for many of today's projects. Literature is now appearing describing PM 2.0, which focuses on new project management tools, better project governance, improved collaboration with stakeholders, and more meaningful information reporting using metrics, key performance indicators (KPIs), and dashboards.

1.1 CHARACTERISTICS OF PM 1.0

Project management had its roots in the aerospace, defense, and construction industries more than 50 years ago. Project management practices were effective on large projects with reasonably known and predictable technology, assumptions, and constraints that were unlikely to change over the duration of the project and a somewhat stable political environment. Unfortunately, for most companies, these types of projects represented only a small portion of all of the projects that companies needed to complete to remain in business.

Today, we are applying the project management approach to a wider variety of projects encompassing all areas of business where politics, risk, value, company image and reputation, goodwill, sustainability, and quality are seen as being potentially more important to the firm than the traditional time, cost, and scope constraints. As such, the traditional project management practices that we have used for decades, which we shall call PM 1.0, are now seen as ineffective for managing some of these new types of projects.

1

PM 1.0 is based upon the following activities:

- Projects are identified, evaluated, and approved without any involvement by project managers.
- Project planning is done by a centralized planning group, which may or may not include the project manager.
- Even though the planners may not fully understand the complexities of the project, the assumption is made that the planners can develop the correct baselines and plans which would remain unchanged for the duration of the project.
- Team members are assigned to the project and expected to perform according to a plan in which they had virtually no input.
- Baselines are established and often approved by senior management without any input from the project team, and again the assumption is made that these baselines will not change over the duration of the project.
- Any deviations from the baselines are seen as variances that need to be corrected to maintain the original plan.
- Project success is defined as meeting the planned baselines; resources and tasks may be continuously realigned to maintain the baselines.
- If scope changes are necessary, there is a tendency to approve only those scope changes where the existing baselines will not change very much.

With PM 1.0, executives were fearful that project managers might begin making decisions that should be made only at the executive levels. Senior management wanted standardization and control in the way that projects were being managed. Project managers were given very little real authority to make decisions. Almost all important decisions were made by the project sponsors. Enterprise project management (EPM) methodologies were created with the mistaken belief that one size fits all. Every project had to follow the EPM methodology because it supported the executives' comfort zones regardless of the ramifications. The EPM methodologies were constructed around rigid policies and procedures. Project status reporting resulted in massive reports and as much as 25% of a project's budget could be consumed by reporting requirements.

Even though a new edition of the **PMBOK® Guide** comes out every four or five years with changes to get us further away from PM 1.0, the **PMBOK® Guide** still contains many of the elements of PM 1.0. It may not be possible, or even practical, to create a single **PMBOK® Guide** that can satisfy those firms that still prefer PM 1.0 and those that have a necessity for PM 2.0.

1.2 OTHER CRITICAL ISSUES WITH PM 1.0

PM 1.0 has worked well for many companies for the types of projects that they traditionally managed. But for other companies there were significant defects with PM 1.0 that needed to be changed. As an example, conventional project and even business planning, as used with PM 1.0, worked on the expectation that managers can predict future outcomes by extrapolating from past results. Planning is often based upon history. But for many new business opportunities and the forthcoming projects this way of planning was often not possible. Experience may be lacking or extrapolating from past experience may be misleading.

A solution to this problem using PM 2.0 is to predict future outcomes based upon assumptions. Some of the assumptions made during the planning process will very likely come true, whereas the outcome of others may very well impact the project to a point where the project should be redirected or even canceled. Project managers may have to test all of the assumptions by developing contingency plans based upon "what-if" scenarios. However, with PM 1.0, the assumptions that appeared in the business case or the project charter were taken as fact and often never challenged. This resulted is a waste of valuable resources.

There were several other PM 1.0 issues that needed to be corrected with PM 2.0. Some of these were:

- Believing that one project management methodology can be applied to every project
- Taking for granted that the constraints and assumptions that are in the business case/charter are correct and need not be tracked
- Trusting that the planning of others, such as a planning department, is always correct and need not be challenged
- Lacking ownership of plans we did not participate in, resulting in lack of commitment to the project
- Working with a structured project plan that does not allow for the creativity of team members
- Not having all necessary information available to the project team
- Working with sponsors and governance committees that do not understand their roles and responsibilities
- Trusting that all of the decisions made by the sponsors or governance committees are the correct decisions
- Believing that implementing project management by executive decree will make it work
- Having no project management culture in the firm
- Believing that a changeover to a project management culture can happen overnight
- Having project management recognized as a part-time addition to one's primary job rather than seen as a career path opportunity
- Not understanding the need for project health checks or how to perform a health check
- Having limited tools to support project management activities
- Having too many projects and not enough qualified resources
- Wasting time on projects that need resources we do not have
- Not having any optimization of resources
- Having no benefit realization plan
- Not understanding how to track benefits or value
- Not working on the projects with the highest value
- Not recognizing the relationship between the project and strategic business objectives
- Believing that if the project fails, we still have an endless stream of customers
- Not having any collaboration with stakeholders

- Reporting project information vertically up the organizational hierarchy rather than accessing information to the whole team
- Preparing all reports in an optimistic manner with the hope that we can correct any problems before management recognizes the truth

Obviously, there are other issues that could be added to the list. But at least we recognize that a valid need exists for PM 2.0.

1.3 PROJECT MANAGEMENT 2.0

The idea for PM 2.0 came primarily from those project managers involved in software development projects where adding version numbers to project management seemed a necessity because of the different tools now being used and different project needs. Over the years, several studies have been conducted to determine the causes of IT project failures. Common failure threads among all of the studies included lack of user involvement early on, poor governance, and isolated decision making. These common threads have identified the need for distributed collaboration on IT projects. From an IT perspective, we can define PM 2.0 using the following formula:

PM 2.0 = PM 1.0 + distributed collaboration

Distributed collaboration is driven by open communication. It thrives on collective intelligence that supports better decision making. Traditional project management favored hierarchical decision making and formalized reporting, whereas PM 2.0 stresses the need for access to information by the entire project team, including the stakeholders and those people that sit on the project governance committee.

Project data will be displayed over mobile devices such as cell phone or tablet screens.

The need for distributed collaboration is quite clear:

- Stakeholders and members of governance committees are expected to make informed decisions rather than just any decisions.
- Informed decision making requires more meaningful metrics.
- The metrics information must be shared rapidly.

Collaboration through formalized reporting can be a very expensive proposition, which is why PM 2.0 focuses heavily upon project management metrics, KPIs, and dashboard reporting systems. This increase in collaboration leads some people to believe that PM 2.0 is "socialized project management."

Agile project management is probably today's primary user of PM 2.0. However, there is criticism that the concepts of PM 2.0, accompanied by the heavy usage of distributed collaboration, cannot be used effectively on some large projects. This criticism may have some merit. There still exists a valid need for PM 1.0, but at the same time there are attempts to blend together the principles of PM 1.0 and PM 2.0.

All new techniques undergo criticism. PM 2.0 is no exception. Some people argue that PM 2.0 is just a variation of traditional project management. Table 1-1 shows many of the differences between PM 2.0 and PM 1.0. When reading over Table 1-1, we must keep in mind that not all projects such as those utilizing an agile project management methodology will necessarily use all of the characteristics shown in the PM 2.0 column. Project managers in the future will be given the freedom of selecting what will work best for them on their project. Rigid methodologies will be replaced by forms, guidelines, templates, and checklists. The project manager will walk through a cafeteria and select from the shelves those elements/activities that best fit a particular project. At the end of the cafeteria line, the project manager, accompanied by the project team, will combine all of the elements/activities into a project playbook specifically designed for a particular client. Client customization will be an essential ingredient of PM 2.0.

PM 2.0 is not a separate project management methodology appropriate for small projects. It is more of a streamlined compilation of many of the practices that were embodied in PM 1.0 to allow for a rapid development process. The streamlining was largely due to advances in Web 2.0 software, and success was achieved when everyone on the project team used the same tools.

Although PM 2.0 has been reasonably successful on small projects, the question still remains as to whether PM 1.0 is better for large projects. The jury has not delivered a verdict yet. But some of the publications that discuss how PM 1.0 and PM 2.0 can be combined offer promise. Perhaps in a few years we will be discussing PM 3.0. Only time will tell.

There are other activities that differentiate PM 2.0 from PM 1.0. However, for the remainder of this book, primarily the PM 2.0 activities in Table 1-1 will be discussed.

TABLE 1-1 Differences between PM 1.0 and PM 2.0		
Factor	**PM 1.0**	**PM 2.0**
Project approval process	Minimal project management involvement	Mandatory project management involvement
Types of projects	Operational	Operational and strategic
Sponsor selection criteria	From funding organization	Business knowledge
Overall project sponsorship	Single-person sponsorship	Committee governance
Planning	Centralized	Decentralized
Project requirements	Well defined	Evolving and flexible
Work breakdown structure (WBS) development	Top down	Bottom up and evolving
Assumptions and constraints	Assumed fixed for duration of the project	Revalidated and revised throughout the project
Benefit realization planning	Optional	Mandatory
Number of constraints	Time, cost, and scope	Competing constraints
Definition of success	Time, cost, and scope	Business value created
Importance of project management	Nice-to-have career path	Strategic competency necessary for success
Scope changes	Minimized	Possibly continuous
Activity work flow	In series	In parallel

(continued)

TABLE 1-1 (*Continued*)

Factor	PM 1.0	PM 2.0
Project management methodologies	Rigid	Flexible
Overall project flexibility	Minimal	Extensive, as needed
Type of control	Centralized	Decentralized
Type of leadership	Authoritarian	Participative (collaborative)
Overall communications	Localized	Everywhere
Access to information	Localized and restricted	Live, unlimited access and globalized
Amount of documentation	Extensive	Minimal
Communication media	Reports	Dashboards
Frequency of metrics measurement	Periodically	Continuously
Role of software	As needed	Mandatory
Software tool complexity	Highly complex tools	Easy-to-use tools
Type of contract	Firm fixed price	Cost reimbursable
Responsibility for success	With project manager	With the team
Decision making	By project manager	By the team
Project health checks	Optional	Mandatory
Type of project team	Colocated	Distributed or virtual
Resource qualifications	Taken for granted	Validated
Team member creativity	Limited	Extensive
Project management culture within firm	Competitive	Cooperative
Access to stakeholders	At selected intervals	Continuous
Stakeholder experience with project management	Optional	Mandatory
Customer involvement	Optional	Mandatory
Organizational project management maturity	Optional	Mandatory
Life-cycle phases	Traditional life-cycle phases	Investment life-cycle phases
Executive's trust in project manager	Low level of trust	High level of trust
Speed of continuous improvement efforts	Slow	Rapid
Project management education	Nice to have but not necessary	Necessary and part of life-long learning

1.4 CRITICISM OF PM 2.0

All new techniques bring with them both advantages and disadvantages. The disadvantages will most certainly undergo criticism. PM 2.0 is no exception. Examples of some of the criticism are:

- Some people argue that PM 2.0 is just a variation of traditional project management and that the changes would have happened anyway.
- Many companies have track records of success using PM 1.0. Asking them to now use PM 2.0 may lead to unnecessary problems.
- PM 2.0 works only on IT projects, especially those requiring use of agile or Scrum techniques.
- PM 2.0 advocates open communications, and this may not be possible on large projects. Distribution and control of proprietary information could be an issue as well.
- The data distributed in PM 2.0 may not be auditable, whereas most people believe that PM 1.0 data are auditable.
- Additional tools will have to be created to support PM 2.0 implementation. The cost of developing the tools may be expensive.
- Data requirements can easily get out of control and we can end up with information overload.
- Although PM 2.0 focuses on collaboration, there is no guarantee that stakeholders or governance committee members will communicate freely with one another.
- Even though PM 2.0 will most certainly benefit strategic as well as operational projects, there is no guarantee that executives will allow project managers to manage strategic projects even if governance is provided.

There are naysayers that will argue against any new technique that may be perceived as pulling them away from their comfort zone. Only time will tell if the criticism has any merit. But one thing is for certain: PM 2.0 is being implemented and it works.

1.5 PROJECT MANAGEMENT 2.0: TECHNOLOGICAL BLESSING OR CURSE?*

PM 2.0: Blessing or curse?

There is no question that advances in technology have impacted and changed both our professional and personal lives in ways that most of us could not have imaged 20 years ago, or maybe even 10 years ago. The combination of mobile access to an ever-expanding Internet has created a level of connectivity to information and remote ideas that is unprecedented in the history of humanity, perhaps second only to William Caxton's first printing of a book in English in 1473.

Today we live and work in a Web 2.0 world of the three dynamic C's: namely, connectivity–context–collaboration.

The smartphones, tablets, and smart technology that we carry with us as we move through our daily lives enable us to be connected not only to online information but also to people. One only has to stand in a public space and look around to see that

---

* Material in this section was graciously provided by John R. Winter, Vice President—Global Learning Solutions, International Institute for Learning, Inc.

9 out of 10 people are talking, texting, checking Facebook, using their smartphone to get directions, or "checking" into their current location using apps like Foursquare or Find My Friends. There is no doubt that the mobile phone has become ubiquitous.

The smart devices we carry with us now are also context savvy; they can pinpoint our location and present information to us that may be helpful or important based on our location. We may be walking by our local drug store when our phone buzzes or sounds an alert to remind us that we need to pick up the prescription we phoned in earlier. The phone's global positioning system (GPS) knows where we are in relation to the store's location and this triggers the app where we set up a reminder earlier.

Context also works in the pull sense too. Try using an app such as Flixster to see where a movie is playing and it will present you with the location and movie times for those playing in theaters around you. So not only are we now connected, but we are connected in a powerful way incorporating the additional layer of context.

Collaboration, sharing, and user-generated content are very much at the heart of Web 2.0. To most of us, Facebook, Twitter, and Instagram are instantly recognizable names and have given rise to a whole generation of Web denizens who are very familiar with sharing information, creating and posting their own user-generated content, as well as collaborating in joint problem solving.

If we so choose, our smart devices will help us keep organized, ensuring we do things in a timely manner and enabling us to gather and share information rapidly among our family, friends, and work colleagues. These activities, it can be argued, are also at the heart of effective project management.

The project manager operating in a Web 2.0 world has this new set of tools at his or her disposal to support the successful completion of the project work in distributed teams. Working in a collaborative and distributed manner is nothing new; we have been using enterprisewide, server-based project management software for many years now. It could be argued that the possibility of working in a collaborative way, with distributed teams, was first brought to our attention by the late Douglas Engelbart in 1951 when he first wrote about this concept and went on to demonstrate it in 1968 along with his other inventions, the computer mouse, networked computers, and the early days of graphical user interfaces, in what was called "The Mother of All Demos."

What is new, of course, is the hardware, software, and far-reaching Internet that we now have available to achieve what back in the late 1960s almost seemed like science fiction. Today we carry in our pockets devices that have infinitely more computing power than that of the Apollo 11 Space Shuttle, which got Armstrong, Aldrin, and Collins safely to the moon and back. Technology writer Grant Robertson wrote of Apple's iPhone in a blog post several years ago, "The iPhone is so advanced compared to the computer used in Apollo's guidance system that it's hard to believe they both came from the same planet—at roughly the same period in time when viewed in contrast to man's time on Earth."[1]

So we now have amazing hardware and a whole ecosystem of software and apps to run on our devices, but is all that sufficient to explain our adoption of the Web 2.0 approach to our work, life, and project management? I would argue it is not. If individuals do not quickly understand what they can achieve with a new piece of technology and

---

[1] "How powerful was the Apollo 11 computer?" Grant Robinson, July 20, 2009.

what it can do for them, that is, how it can help make their lives better/easier, they will quickly turn away and look for the next innovation.

One only has to think back to Apple's Newton that was released in 1993 and was subsequently discontinued in February 1998. This product was really the precursor of what is today a tablet, or tablet computer. Some people still maintain it was an amazing device that was ahead of its time, which indeed it probably was. However, it never achieved the kind of mass adoption that smartphones and tablets have today, because people could not see how it made life better for them and it ended up being coveted by "geeks" and "tech-heads"!

Simplicity and ease of use also contributed to our willingness to embrace the Web 2.0 world. Even though a staggering amount of complex engineering and programming is taking place just behind the glass surface of our touch-screen devices, what we appreciate most is the fact that it appears simple to use and just works in a reliable and predictable way.

Much like our day-to-day life, the work life of busy project, program, or portfolio managers is complex and there is a real need for the powerful handheld devices to help sift through, control, and coordinate the massive amounts of information that assault their senses every minute of every day.

If you have not come across the acronym VUCA yet, you probably will hear it soon. It comes from the U.S. Army training for soldiers and leaders of soldiers in intense combat situations. It describes situations they face which are volatile–uncertain–complex–ambiguous and how to account for this when gathering information, making decisions, and providing direction.

This categorization of factors can easily apply to situations in which project managers find themselves having to achieve results. Although the consequences may not be life or death as they are in the military, nonnegotiable changes in scope; reallocation of resources; unexpected cuts in budgets; midproject changes in stakeholders or the project sponsor; multiple schedule disruptions; changes in project deliverables; leading global team members with distinct cultural biases; all can at times make it feel like a VUCA battlefield for the project manager.

Successful project managers (of PM 2.0) not only need to understand effective project management methodology and be skilled in leading their project teams but also must know how to manage fast-paced and multisourced flows of information, curate it, and make the right decisions in a timely manner. Today that means using the hardware and software of today's Web 2.0 environment.

The key to being an effective PM 2.0 mamager is mastering the art of being connected and using the best technology tools to get the job done in a collaborative way while working with a distributed team.

The importance of managing connectivity is probably more critical during the planning, executing, and monitoring and controlling phases of a project, that is, when the bulk of the project work is being completed. However, I am not discounting the importance of initiating or closing, as giving these phases scant focus and attention will invariably lead to major project troubles. But missing a vital piece of information or a warning sign when the project engine is running at full throttle and it is full steam ahead can quickly jeopardize a project's success.

So how does a PM 2.0 manager use the technology? What is available now? What might we expect in the future? I hate to provide this clichéd response, but it depends. It really does. It depends on the size of the project; the way in which the team is distributed; organizational readiness in terms of the technological adoption, knowledge, and cultural acceptance; and the technological development in general.

So with this in mind I am not attempting to address each of these factors but instead assume that the ideal situation exists for what follows.

Much of the success of the PM 2.0 manager rests on the software and hardware that he or she has in place to control the flow of incoming project information; process and make decisions based on the information; and communicate out to project team members, stakeholders, and clients.

In the Facebook/Twitter society in which we live there is already a great familiarity with the type of social media software necessary for the management of PM 2.0 information. While Facebook and Twitter would not necessarily be considered appropriate tools by chief information officers (CIOs), there are many alternative enterprise software solutions available that will achieve the same result safely behind an organization's firewall. Probably one of the best known tools is Yammer (http://www.yammer.com/), a private social network software that helps employees collaborate across departments and locations. A tool such as this, with its wiki-like format, encourages the sharing of informal information and joint problem solving.

"The devil is in the details," as the expression goes, and it is not uncommon for important details within the execution and tracking of a project to go overlooked, sometimes with serious consequences. As project complexity increases, the likelihood of this increases too. With a forum to pose questions, raise concerns, and get answers, missing the detail because it is buried in the in-box of someone's e-mail is less likely.

Yammer of course is not the only solution, and large software companies such as Microsoft and Google have many cloud-based enterprise resource planning (ERP) tools for large organizations that can be used for this purpose. In some respects the tool itself is almost irrelevant; what is important is that an electronic forum exists where project information can be shared informally, in addition to formalized dashboards, especially when working with a distributed team.

Here is an example of how Yammer or a similar tool can be used to facilitate the informal flow of project information. Within Yammer a main group can be established for a specific project and within the main group subgroups can be created for project team members, stakeholders, and so on. People connected with the project can post comments, questions, ideas, and suggestions in either the main group, which will be seen by everyone who belongs to the organization's Yammer group, or just to the specific people within their subgroup.

This is how it might work. Omar, who is a team member of a large engineering project in the north of England, has spotted an odd phenomenon while visiting an oil drilling rig in the North Sea. He is a little concerned because he notices that when the catalyst for an ultrastrong adhesive is added to a compound that will be used to hold steel plates together, it is causing the adhesive to harden much faster than usual. He wonders whether this will cause a problem with the integrity of the bond between the plates in the future. It is something he has never experienced before and neither has any of the three other teams members who are working with him on the rig.

Omar has a nagging doubt that this may be significant, but he does not want to stop the project, which is already behind schedule for something that may be nothing at all.

He knows that other members of his project team are working on similar projects in other parts of the world, so he logs onto his company's social enterprise network that has been set up for the project and posts a description of his observation along with some photographs he has taken showing different phases of the engineering process.

As soon as his post is made, Elvin, the project manager in Singapore, feels a buzz from the smartphone in his pocket. Although he is having dinner with friends, he excuses himself, steps out of the restaurant, and checks his phone. The icon on his social enterprise network app shows five messages, then immediately jumps to six, then seven, and finally stops at nine. He quickly taps the app and reads the posts. He sees Omar's original message and the responses from other project team members in Argentina, Australia, and India. All three agree that this is out of the ordinary and Sumita in India remembers a similar incident several years before when the integrity of the bond between the plates eventually failed. It was due to a faulty batch of chemical compounds.

Elvin immediately posts his response to Omar, instructing him to temporarily halt the current work and then forwards Omar's post to the technical support of the adhesive, who is also part of the main group for the project on the company's social enterprise network.

The technical support representative immediate confers with the chemical engineers within his company's own social network and, after a brief Skype video conference with Omar, a working solution is provided, averting what could have been a very costly future structural failure. This allowed the project to proceed with the minimum of disruption to the schedule.

While this is a fictitious story, it clearly demonstrates the need for the PM 2.0 manager to be connected within context and be able to collaborate quickly and effectively across time zones and geographical boundaries the minute important situations arise.

Apart from social enterprise networks running behind organizational firewalls, as in the previous example, there has also been a rapid growth in new project management software and apps that will run on laptops, smartphones, and tablets. Initially these apps were designed to report, track, and monitor project activity and status in real time, but now they have been developed to incorporate the social network elements vital to the PM 2.0.

Cloud-based solutions such as **AffinityLive** (http://www.affinitylive.com/products/projects), **PieMatrix** (http://www.piematrix.com/), **Box** (http://www.box.com/business/project-management), and **Deskaway** (http://www.deskaway.com/) are a new breed of Web 2.0 project management software built specifically for small- to medium-sized projects that are designed to connect people with information within context and enable online collaboration.

You may wonder what is next. When we begin to venture into the world of the "thinking internet," the semantic web, or Web 3.0, what will life be like for the project manager then?

We know already that we are rapidly entering the age of wearable technology, with developments like Google Glass, although still not mainstream, and gadgets that measure our activity, such as the Fitbit, Jambox's Up Band, and Samsung's Gear Fit, which are definitely mainstream. We are also just beginning to witness clothing with the technology woven directly into the fabric of the garment.

So linking this with the fact that the era of the "Internet of Things" is slowly dawning where objects previously not connected are beginning to be connected wirelessly to the Internet—yes your refrigerator can have its own Internet address! The future is beginning to look interesting.

It is easy to foresee that PM 3.0 manager may be sitting at home watching a movie, when suddenly the light in the table lamp beside her starts to pulse slowly, or turns to a deep magenta color, having received a message from the project management software on her tablet signaling the status of one of her projects has just changed from yellow to red.

Or, sitting in a theater during the performance of a play, the fabric in the right sleeve of another project manager's shirt begins to quietly vibrate, discretely informing him that he needs to look at his smartphone during the intermission to attend to an issue that has arisen on one of his projects. And, yes, the smartphone knew he was in a situation where he needed to be notified graciously. And, yes, it knew which play he was watching and when the intermission would take place and had calculated that based on the severity of the issue the next 20 minutes would find.

Whether these scenarios ever become a reality or not—I suspect they will—the important thing is for managers of projects to strive to be connected to the information they need, within the context of their current situation, so they can collaborate without boundaries to their distributed teams. How they will do this and what technology will help them present a future full of exciting possibilities.

1.6 POLICING PM 2.0

Who is responsible for policing PM 2.0?

Copyright © Scott Maxwell/Fotolia

It is wishful thinking to believe that all of the PM 2.0 activities listed in Table 1-1 will evolve naturally. Some of the changes may be initiated by senior management, others by functional management, but most of them will be the result of project management initiatives. Someone must assume responsibility for the policing of the changeover from PM 1.0 to PM 2.0 and making sure that the transition goes smoothly. Without some sort of structure and guidance, the initiatives can take much longer than necessary, which will then prolong the time needed to see the benefits of PM 2.0. The policing function must be performed by the project management office (PMO).

Traditionally, PMOs were created to help promote the installation and growth of project management. This included creating a project management methodology and the accompanying forms, guidelines, templates, and checklists. As the number of project management successes increased, management began assigning additional responsibilities to the PMO. Some of these responsibilities are:

- Forms for standardization in estimating
- Forms for standardization in planning

- Forms for standardization in scheduling
- Forms for standardization in control
- Forms for standardization in reporting
- Clarification of the project manager's roles and responsibilities
- Preparation of job descriptions for project managers
- Preparation of archive data on lessons learned
- Continuous project management benchmarking
- Developing project management templates
- Developing the project management methodology
- Recommending and implementing changes and improvements to the existing project management methodology
- Identifying project management standards
- Identifying best practices in project management
- Performing strategic planning for project management
- Establishing a project management problem-solving hotline
- Coordinating and/or conducting project management training programs
- Transferring knowledge through coaching and mentorship
- Developing a corporate resource capacity/utilization plan
- Assessing risks in projects
- Planning for disaster recovery in projects
- Performing or participating in the portfolio management of projects
- Acting as the guardian for project management intellectual property

Companies began recognizing the return on investment of using a PMO. It is therefore a natural follow-on for the PMO to take the lead with PM 2.0 implementation activities. However, there are significant challenges. Perhaps the greatest challenge is that PM 2.0 is now aligned to strategic business objectives as well as the operational objectives most commonly used with PM 1.0. The PMO must now monitor closely how PM 2.0 will interface with all business units rather than just those functional areas that are using project management.

1.7 WORKING WITH STAKEHOLDERS IN PM 2.0

From the birth of project management in the early 1960s up to the last decade, stakeholder involvement in projects has been more passive than active. Stakeholders focused heavily on the deliverables at the end of the project. And, if they did get actively involved at all, it was close to the end of the project when there were fewer decisions for them to make.

During this time period, stakeholders knew very little about the actual processes used in project management. This included internal stakeholders, stakeholders from the client's organization, and governance committee stakeholders. Everything was end-results oriented. Information provided by the project manager was considered as the Gospel, never questioned, and the stakeholders had no way of validating whether or not this was the right information. When decisions had to be made, it was most often seat-of-the-pants decision making rather than informed decision making based upon meaningful information. Simply stated, stakeholders did not know what information they needed and focused mainly on just time and cost metrics.

Today's View of Stakeholder Relations Management

Today, stakeholders appear to be much more knowledgeable about project management than in the past. Stakeholder involvement is much more active than passive, and the involvement begins right at the initiation of the project. Continuous stakeholder involvement is mandatory rather than optional as indicated in Table 1-1 as a primary characteristic of PM 2.0. There are several driving forces which necessitated this change:

- The projects we are working on now are more complex than in the past.
- Complex projects most often have a higher degree of risk associated with them.
- Stakeholders are expected to be and want to be more actively involved in certain critical decisions.
- Stakeholder involvement in project risk management requires meaningful information.
- Stakeholders understand the difference between traditional decision making and informed decision making necessary for a PM 2.0 environment.
- Stakeholders want to participate in the decision regarding what metrics they wish to see in order to monitor project progress.

As stakeholder involvement became more active than passive, project managers soon realized that that the way that they handled stakeholder relations management also had to change. Project managers must now:

- Work closely with all of the stakeholders to understand the requirements of the project rather than relying solely upon the client for requirements definition.
- Work closely with each stakeholder or stakeholder group to understand what metrics they wish to have reported and how frequently.
- If necessary, create a separate project management information system for each stakeholder.
- Be aware that the information system will report status in a dashboard format, and there may be a different dashboard for each stakeholder.
- Have a dashboard designer as part of each project team.
- Understand that stakeholders now recognize the importance of informed decision making rather than ordinary decision making based upon guesses.

The latest version of the **PMBOK® Guide—Fifth Edition** introduced a new knowledge area, namely stakeholder management. In my opinion, it would have been better to call the new area of knowledge stakeholder relations management because project managers do not manage the stakeholders. Project managers may have some control over managing the relationships, but not managing the actual people. Most of the stakeholders may very well be at a much higher position in their respective organizational hierarchy than the project manager.

The starting point in managing stakeholder relations is a clear understanding of what is expected from the stakeholders in the way of authority, responsibility, and decision making. We traditionally map the stakeholders on a power–influence grid and provide most of our attention to those stakeholders that have a great deal of power and can influence the direction of the project. Today, the stakeholders in this quadrant of the grid, possibly along with all of the other stakeholders, are expected to assist the project manager by making informed decisions. Making informed decisions requires that correct and meaningful metric information be presented to them in a timely manner.

Need for Meaningful Information

For years, stakeholders never fully understood metrics. They knew that a metric was a measurement, but they often failed to understand that not all metrics are equal in importance and that not all metrics provide meaningful information for decision making. Today, we differentiate between metrics and KPIs. Key performance indicators are those critical metrics that substantiate the health of the project and can be used to predict the future success or failure of the project. Project managers can identify up to 50 metrics on projects but usually somewhere between 8 and 10 metrics are considered KPIs. The KPIs are what stakeholders need to see for informed decision making.

All That Glitters Is Not Gold

Providing stakeholders and governance committee members with PM 2.0 metric/KPI information is certainly the correct thing to do. However, when something appears to be a great idea, there are always opportunities for bad things to happen.

PM 2.0 metrics management issues can create severe problems when dealing with stakeholders or members of a governance committee. Some of the more critical issues that may surface are:

- *What happens when stakeholders become infatuated with metrics and want all of the metrics in your metric library displayed on the dashboards?* If you have 50 metrics in your library, you will end up providing too much information to the point where you have information overload. The dashboard viewers may not be able to recognize which metrics/KPIs are critical for informed decision making. This could slow down the decision-making process rather than speeding it up.
- *What happens when stakeholders request specific metrics that you do not understand and do not have the organizational process assets to perform measurements?* This could cause delays in the execution of the project as well as delays in decision making. The project team may need to be trained in how to perform new types of measurements for client-specific metric requests.
- *What happens when stakeholders have disagreements with what the metric data show and conflicts will then occur?* This can happen even with the best dashboard designs.
- *What happens when stakeholders state that they do not want to hear any bad news or see bad news displayed on the dashboards?* This could eliminate effective stakeholder support during a crisis.
- *What happens when stakeholders want to see the data before it appears on the dashboards and filter the information such that they end up stretching the truth?* This could be seen as a violation to project managers' code of professional conduct.

Obviously, metric management does have a down side. But there are approaches that can be taken to minimize the risks, as will be discussed in later chapters.

1.8 FINDING THE INFORMATION

There are numerous benefits to using PM 2.0. Most of the benefits are derived from overcoming the challenges imposed upon us from using PM 1.0. One of the biggest challenges with PM 1.0 was the inability to find enough performance information to determine the true health of the project. We relied heavily upon time and cost as the two primary metrics for measuring and reporting project health because they were the easiest to track. Unfortunately, time and cost alone cannot determine the true health or status. This was known quite well in the early years of project management, but metric measurement techniques were just in the infancy stage. Therefore, only time and cost were used because they were the easiest to measure and report.

How easy will it be to find status information with PM 2.0?

With PM 1.0, computer technology was in the infancy stages and the only software that was readily available was software associated with the earned value measurement system (EVMS). Status reports were printed out monthly and included direct labor, indirect labor (i.e., overhead rates), procurement costs, and other incidental costs such as the use of consultants, travel, printing, training, and conferences. Some companies were able to report weekly status, but for direct labor only. This approach unfortunately showed that any significant crisis may not be known in detail until the next monthly report appeared. Valuable time was lost when effective decisions could have been made.

We learned with PM 1.0 that the true status cannot be determined from just one or two metrics. It is possible that success could be measured by one metric, such as customer satisfaction or the number of deliverables provided to the client at the completion of the project. However, these situations are far from the norm. There are companies that have been successful with PM 1.0 and will probably continue using PM 1.0 in the near term.

In PM 2.0, we will be working on strategic as well as operational projects. Some of these projects may last for 10 years or longer, require hundreds of employees, go through numerous scope changes, and have the membership of governance committees change several times over the life of the project. The complexity of the projects will increase as well, thus mandating more metrics than used with PM 1.0.

The situation becomes even more complicated when project managers are asked to make business as well as project decisions. Significantly more metrics, especially business metrics, will be used with PM 2.0 than with PM 1.0. Most project managers may be unfamiliar with all of the business metrics that companies are using and how they interface with project-oriented metrics. Complexity will occur if the metrics show that what is in the best interest of the business is not in the best interest of the project, or vice versa. With PM 2.0, decision making will thrive based upon the abundance of meaningful metrics. However, as stated in Section 1.4, care must be taken to prevent the use of the new metrics from bringing forth additional problems.

1.9 PERCENT COMPLETE DILEMMA

The EVMS, which will be discussed in later chapters, focuses heavily on just time and cost metrics. On some projects, we may be able to approximate the status of the project from just time and cost as long as we know the performance percent complete with reasonable accuracy. But knowing percent complete is just a guess. Functional managers generally report back to the project manager their best guess on percent complete for the work performed in their functional areas. The situation can get complicated if several functional managers are working on the project and they all have a different opinion of the percent complete.

In the EVMS, the most important term is EV, which is the earned value, or the amount of the work performed to date, expressed in hours or dollars. The simplest equation to calculate EV is

$$EV = \text{percent complete} \times BAC$$

where BAC is the budget established for the completion of the project. If we are unable to determine percent complete with some degree of accuracy, then we may be providing the client with inaccurate information. To alleviate this problem, formulas were created to crudely calculate EV without ever having to accurately determine percent complete. The cost management section of most project management textbooks describes the use of these formulas. Some of the formulas are:

How accurately can we calculate percent complete?

Copyright © Scott Maxwell/Fotolia

- **50/50:** Half of the budget is earned for each element and recorded at the time that the work is scheduled to begin, and the other half at the time that the work is scheduled to be completed.
- **0/100:** Usually limited to work packages (activities) of small duration (i.e., less than one month). No value is earned until the activity is complete.
- **Milestone:** This is used for long work packages with associated interim milestones, or a functional group of activities with a milestone established at identified control points. Value is earned when the milestone is completed. In these cases, a budget is assigned to the milestone rather than the work packages.
- **Percent Complete:** Usually invoked for long-duration work packages (i.e., three months or more) where milestones cannot be identified. The value earned would be the reported percent of the budget.
- **Equivalent Units:** Used for multiple similar-unit work packages, where earnings are on completed units, rather than labor.
- **Cost Formula (80/20):** A variation of percent complete for long-duration work packages.
- **Level of Effort:** This method is based on the passage of time, often used for supervision and management work packages. The value earned is based on time expended over total scheduled time. It is measured in terms of resources consumed over a given period of time and does not result in a final product.

With PM 2.0 and the accompanying growth in metrics, we may find it easier to determine percent complete or at least improve our approximation of percent complete. However, finding the data we need may prove difficult. Project teams will have to perform "data-mining" activities, which will include the identification of new metrics, new measurement techniques, and better performance reporting. Finding all of this information will not be easy, but progress is being made and the benefits are rewarding. Significantly more information is needed for performance reporting with PM 2.0 than with PM 1.0, and many companies have already started creating PM 2.0 metrics.

1.10 INFORMATION OVERLOAD

There's an old saying, "Be careful what you wish for because you may get it!" As with any new technique, people often go to extremes rather than following the straight and narrow or simplest path. The real fear with the quest for metrics is when a "metric mania" mentality sinks in and people look for the maximum number of metrics that can be collected rather than just what is needed.

While this approach of collecting more metrics than needed may have some merit, the result of all of this is usually information overload. The real fear is that everyone will want the metrics they found to be permanently part of the metrics database. Not all metrics carry with them an informational value that justifies their use. People may end up collecting metrics without fully understanding what the metric really means or how it should be used. As will be shown later in this book, simple metrics like time and cost can mean different things to different people.

Can information overload occur in PM 2.0?

Copyright © Scott Maxwell/Fotolia

When information overload occurs, it may become difficult to identify a core set of metrics for the project. Providing clients and stakeholders with too much or too little information can slow down a project. One of the responsibilities of the PMO's policing activities with PM 2.0 is to ensure that the correct metrics are placed in the metric library.

1.11 CUSTOMER SATISFACTION HEADACHE

People seem to be enamored with the belief that customer satisfaction can be obtained from information overload, regardless of whether we are discussing an external or internal customer. Assume that metric mania sets in and you collect significantly more metrics than you need. All of the metrics are placed in the metric library. You just won a contract through a competitive bidding process and the firm-fixed-price contract has been signed. At the beginning of a project, you ask your external customer what metrics they would like to see on their project dashboards. You show the customer your metric library, and the customer then says that they would like to see all of the metrics reported on their dashboards. To make matters worse, they would like to see the dashboards in real time. While this may lead to customer satisfaction, your may have just created a migraine headache for yourself.

Can a large metric library cause headaches?

Copyright © Scott Maxwell/Fotolia

For companies that survive on competitive bidding, there can be a very large cost associated with the identification, collection, tracking, measuring, and reporting of a large number of metrics. During competitive bidding, you may have assumed that you would provide your customer with just one dashboard containing 6–10 critical metrics to track the project. This is what you priced out. After contract go-ahead, the customer sees all of the metrics in your library and wants them all reported. Unless you are able to push through a contract modification to account for the cost associated with the additional metrics, the project may absorb a financial hit.

The costs can be equally as bad when managing a project for an internal client. Even internal clients may ask for more metrics than they need. Giving clients too many metrics is an invitation for client micromanagement.

All metrics age and therefore periodic reassessment of the ongoing value of using each metric must be made. Maintaining a large metric library may not be cost effective and may result in migraine headaches. A possible remedy for the customer satisfaction headache is to prepare a list of recommended metrics that you believe should be used on the project. Allowing the client to make the decision could be a serious mistake. As we develop a history of metrics used and continuous improvements on the metrics, it should be easier for a project manager to convince customers on what metrics should be used. But, once again, the uniqueness of each project may cause headaches initially.

With PM 2.0, it may not be possible to regulate the number of metrics contained in the metric library. Project managers will be making both project and business decisions in PM 2.0. They will be expected to use both project and business metrics when discussing project status. Therefore, metric libraries will contain an abundance of business- and project-related metrics. It is entirely possible that, as the metrics library grows, all of the metric libraries and best practices libraries will be replaced by a single knowledge management system which will include:

- Project metric libraries
- Business metric libraries
- Best practices libraries
- Specialized knowledge libraries
- Benchmarking activities
- Continuous improvement activities
- Other knowledge repositories (i.e., historical project failure analysis data)
- Databases (e.g., estimating databases, client information databases)

1.12 DETERMINING PROJECT HEALTH

With PM 1.0, status reporting was based upon just two primary metrics, time and cost. While it is true that we did look at other metrics in PM 1.0, the EVMS that was created focused heavily upon manipulations of just the time and cost metrics. We had a lack of understanding concerning the measurement techniques that could be used to track other metrics. Status reporting was often more of a guess than based upon fact. The result was a relatively poor understanding of the health of the project.

Project status was calculated primarily from time cards that indicated the hours spent on a work package. The hours were converted to dollars using either the actual salary of the workers or a departmental blended labor rate for a particular pay grade. Actual percent complete was difficult to estimate and therefore snapshots of work-in-progress were considered unnecessary. Customers often did not know the status of their project until the project was completed.

As stated previously, PM 2.0 projects are generally more complex and costlier than PM 1.0 projects. Waiting until we get close to the end of the project to determine the true status will not satisfy most of today's clients or members of the governance committees. Fortunately, today we have more sophisticated software that allows us to not only track

How many snapshots are needed to determine the true health of a project?

dozens of metrics at once but also report the information in real time. Dashboards can be updated as fast as the information can be inputted into an Excel spreadsheet. Therefore, the benefits with PM 2.0 are an infinite number of status snapshots in real time. This allows decision makers and members of the governance committee to make informed decisions rather than seat-of-the-pants decisions based upon a guess.

While real-time metrics display status, they may not clearly indicate the root cause of a problem. They may show only surface conditions. As an example, snapshots of time and cost may indicate that the project is running late and over budget. The project manager may have to dig deeper than just these metrics to find the actual cause of the problem. The problem may be caused by poor workmanship, a degradation in quality, unresolved action items that are causing delays, or a lack of resources. While having surface metrics is seen as a necessity, there is also a need for subsurface metrics as well.

The words *health* and *status* have been used interchangeably in this section. There is a difference between them, and this can best be described by looking at the four types of performance reports that were traditionally prepared with PM 1.0:

Progress Reports: These reports indicate the physical progress to date, namely, how much work was scheduled up to this point in time, how much work was actually accomplished, and how much money was spent. The report might also include information on material procurement, delivery, and usage, but most companies have separate reports on procurement of materials.

Status Reports: These reports identify where we are today and use the information from the progress reports to calculate variances or deviations from the project plan.

Projection Reports: These reports calculate forward-looking projections based upon trends. These reports emphasize where we will end up.

Exception Reports: These reports identify exceptions, problems, or situations that exceed the threshold limits on such items as variances, cash flow, resources assigned, and other such topics.

When we take snapshots of a project, we are collecting data related to how much progress has been made. Snapshot information goes directly into progress reports. The project team then takes the progress data and compares it to the previous reporting period to create the status (or variance) reports. Assumptions may be made as to the reasons for the variances. The projection reports extrapolate the status information into the future, and once again assumptions may be made. The less the number of assumptions made, the more confidence the reader has of the information that would traditionally appear in each of the four mentioned reports. The more metrics you use, the less the number of assumptions that must be made. With PM 2.0 and the use of additional metrics, dashboard performance reporting is expected to provide a more accurate picture of the project's health.

1.13 DASHBOARD RULES FOR DISPLAYING DATA

Having a metrics library with an abundance of metrics may provide no useful benefit unless the information can be properly displayed in such a manner that it can be easily understood. There are rules that most dashboard designers follow. They include:

Rules for Selecting Right Artwork: There are several images that can be used for each metric. Some images may be inappropriate. For example, gauges should not be used for displaying trends.

Rules for Screen Real Estate: There is only so much space available on a computer screen for images. Usually, only 6–10 images should be displayed on a screen.

Rules for Artwork Placement: Some people believe that the most important image belongs in the upper left corner of the screen whereas others believe it should be displayed in the upper right corner.

Rules for Color Selection: Usually the softer colors are used for metrics. Brighter colors are used to highlight critical pieces of information. There are also colors that should be used for people that are vision impaired.

Are there guidelines for displaying metrics on dashboards?

Copyright © Scott Maxwell/Fotolia

Rules for Accuracy of Information (2D vs. 3D): While three-dimensional graphics looks impressive, there may be some difficulty with accurately reading the data. Most graphic designers focus on two-dimensional graphics.

Rules for Aesthetics: The display of the graphics must be pleasing to the eye.

If the viewers of the dashboards cannot understand what they are seeing, they may lose faith in the entire dashboard concept. This could lead to devastating results. Some companies prefer having pilot courses for first-time viewers to make sure that they understand what they are seeing. Because the space is limited on dashboards, care must be taken to avoid the heavy usage of company or project logos and other branding information. Company branding is always nice to have, but screen real estate is limited and expensive. Cluttering up a dashboard with too much information can lead to information overload.

1.14 REDUCTION IN COST OF PAPERWORK

PM 1.0 thrived on written reports. In some cases, the reports accounted for between 25 and 50% of the project's budget. Steps normally included in report preparation are:

- Organizing
- Writing
- Typing
- Editing
- Retyping
- Proofing
- Graphic arts
- Approvals
- Reproduction
- Distribution
- Storage
- Disposal

Can PM 2.0 get us closer to "paperless" project management practices?

Copyright © Scott Maxwell/LuMax-Art/Shutterstock

Typically, the cost per page for a report provided to a client runs between $1200 and $2000 based upon 8–10 hours per page for everyone involved in the above steps and using a fully burdened hourly labor rate. In some estimates, workers may spend as much as 25% of their time writing reports. And to make matters worse, it is entirely possible that the reports are never read.

With PM 1.0, we are often plagued with staffing the project with people who have writing skills if we know in advance that reports are needed. We may know people who have the technical skills that we would like on a project but we cannot use them because of their poor writing skills. Several years ago, an engineering company selected project managers based almost entirely upon their writing skills.

With PM 2.0, reports are replaced with dashboards that show the most critical metrics on the project. Dashboard viewers see the most critical metrics needed for informed decision making and can connect to other dashboards using drill down buttons if additional information is required. Perhaps the most important benefit is that each dashboard can be customized for the individual viewer rather than giving everyone a massive report.

PM 1.0 supported the need for massive reports, many of which were never read. Because project sponsors and decision makers lacked all of the necessary information for informed decision making, many decisions were delayed and thus increased the cost of the project. An effective metrics management program, spearheaded by senior management, will allow for cost savings on projects.

Can PM 2.0 metrics save us money?

Another area for cost savings is that the project report writers can now spend more time working directly on project activities that are part of project execution rather than writing reports. Not all project team members have writing skills. Not all reports can be eliminated. Those that can be replaced with metrics and a dashboard reporting system will result in cost savings. For companies that survive on competitive bidding, dashboard reporting systems may allow for the submission of a lower bid, thus increasing the chance of contract award.

With PM 2.0, the use of additional metrics and KPIs combined with Web 2.0 technology can save as much as 20% of a project's budget. Although support statistical data do not exist at this time, significant cost savings is expected.

Cost saving does not necessarily mean additional profits. Cost savings can allow the scope of the project to increase without having to add additional funds to the project's cost baseline. We will be doing more work for the same amount of money. Cost savings can allow the portfolio selection committee to work on more projects.

1.15 REDUCTION IN EXECUTIVE MEDDLING

Many of us that have lived with PM 1.0 can attest to continuous executive meddling on some projects. Meddling occurs most often because senior management does not have a clear picture as to what is happening on the project. Once again, time and cost metrics alone cannot provide a clear picture.

Another reason for meddling is that executives may believe that they have something to lose if the project were to fail. This may include:

- Viewing project failure as damage to their career
- Viewing project failure as damaging their reputation
- Viewing a lack of project knowledge as a sign of weakness

- Fear of exposing to others some bad decisions they may have made on the project
- Having to answer questions from stakeholders and not having the necessary information
- Viewing information as power and needing to know as much as possible about the project.

Meddling most often occurs when progress is less than expected or when serious problems occur. Meddling may not be bad unless the executives overreact and try to take over the project. In this case, the project manager is treated like a puppet.

With PM 2.0 and the use of a dashboard reporting system, executives can have daily updates on the status of the project. The dashboards can be customized for the needs of each executive. The need for continuous meddling should be reduced.

Can metrics prevent executive meddling?

Copyright © Scott Maxwell/
LuMaxArt/Shutterstock

1.16 PROJECT MANAGEMENT SKILLS

Project managers historically came from the engineering ranks of a company. The only criteria to becoming a project manager were a command of technology and some writing skills. Technical decisions were made by the project managers, but all business decisions were made by the project sponsors.

Most project managers were never trained on human relations management and lacked the necessary skills to resolve human relations issues and conflicts. On some large project teams, an assistant project manager or counselor was responsible for organizational development issues. Most of the engineers had not taken courses in interpersonal skills, leadership, mentorship, facilitation, or conflict resolution. The counselor assisted the project manager with all behavioral issues.

At that time, project management was in the infancy stages and we were not sure what skills an effective project manager should possess. Not very many companies had job descriptions for project managers. There were no **PMBOK® Guides** on project management or college or university coursework on project management other than possibly in civil engineering programs.

What skills do project managers need?

Copyright © Scott Maxwell/
LuMaxArt/Shutterstock

Today, there are numerous training programs for project managers. At the onset of a project, the project manager may be placed under a microscope to see if he or she possesses the necessary skills for the project. Specialized training may be necessary. Job descriptions are being replaced with competency models which identify the specific skills that a project manager must possess.

With PM 2.0, we are better able to match the correct person to the needs of the project. We have people with skills needed to manage specialized projects, such as a recovery project manager (RPM), who is an expert in turning around failing projects.

1.17 CONTINGENCY PLANNING

With PM 1.0, contingency planning was done sporadically. Projects were allowed to slip and cost baselines were allowed to overrun. To make matters worse, most project

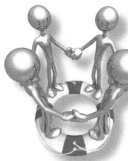

Is there a need for contingency planning?

managers who came from the engineering ranks were highly optimistic in their belief that whatever plan they laid out initially would work successfully and contingency plans were not needed. Contingency plans, if prepared at all, would be developed after a crisis occurred.

One of the reasons for poor contingency planning was a poor understanding of risk management. With PM 1.0, there was a tendency to look at only financial and scheduling risks. Today, with PM 2.0, risk management is maturing and we look at all types of risk.

With PM 2.0, there is significantly more information available to the project managers at project initiation as a result of a better portfolio selection process. Business cases are better defined, knowledge of the skill levels of the resources needed are known beforehand, organizational capacity planning models exist, and we have members on the governance committee that are more knowledgeable in project management.

All of this results in the benefit of being able to develop contingency plans throughout the life of the project. Metric and KPI data provided on a dashboard reporting system may eliminate the previous need for painful data mining to find the necessary information for contingency planning. Members of governance committees are more willing to participate. The need for contingency planning may be reduced, but it will not be eliminated.

DISCUSSION QUESTIONS

The discussion questions are for classroom use to stimulate group thinking about PM 2.0. There are no right or wrong answers to most of the questions.

1. What type of industries, companies, or projects could still be reasonably successful using PM 1.0 rather than PM 2.0?
2. What are the pros and cons of allowing executives to assume the responsibility for policing PM 2.0 implementation?
3. Why is the estimation of percent complete so difficult? Are there situations where it can be estimated with reasonable accuracy?
4. Which people will most likely be the source for creating information overload?
5. Can good metrics eliminate or reduce executive meddling? If not, then how can it be reduced?
6. Who has the prime responsibility to determine if a project health check is necessary?
7. Can PM 2.0 become more costly than PM 1.0 and, if so, under what circumstances?
8. Should the need for contingency planning with PM 2.0 be greater or less than with PM 1.0?
9. Who makes the final decision as to how many metrics should be displayed on a dashboard?
10. Why is stakeholder involvement so important with PM 2.0?

CHAPTER 2

A PEEK INTO THE FUTURE OF PROJECT MANAGEMENT

2.0 CHANGING TIMES

For more than 50 years, relatively modern project management has been in use but perhaps not on a worldwide or companywide basis. What differentiated companies in the early years was whether or not they used project management, not how well they used it. Today, almost every company uses project management and the differentiation is whether they are simply good at project management or whether they truly excel at project management. The difference between using project management and being good at project management is relatively small and most companies can become good at project management in a relatively short time period, especially if they have executive-level support. But the difference between being good and excelling at project management is quite large. The recognition of the need for continuous improvement is critical. The acceptance of PM 2.0 cannot guarantee that the firm will excel at project management, but it will certainly improve the chances of success.

Companies such as IBM, Microsoft, Siemens, Hewlett-Packard (HP), and Deloitte, just to name a few, have come to the realization that they must excel at project management. IBM has more than 300,000 employees with more that 70% outside of the United States. This includes some 46,000 employees managing projects part time or full time. HP has more than 8000 project managers and 3500 PMP®s. HP desires 8000 project managers and 8000 PMP®s. These companies thrive on continuous improvement efforts.

2.1 IMPACT OF RECESSIONS

When a firm's economic conditions are healthy, executives are reluctant to "rock the boat" and implement changes that they believe may be unnecessary at that time. But when things are going poorly and there is a lot of red ink on the firm's balance sheet, companies often embrace project management practices. Table 2-1 shows the impact of the last three recessions. The last column shows how project management had changed.

TABLE 2-1 Impact of Recessions

Recession or Poor Economy	Layoffs	R&D	Training	Solutions Sought	Results of Recession or Poor Economic Conditions
1979–1983	Blue collar	Reduced or eliminated	Reduced or eliminated	Short term	Return to the status quo No real project management support No allies for project management
1989–1993	White collar	Reduced or eliminated	Reduced or eliminated	Long term	Project management support growth Project management becomes a career path Growth in importance of risk management Change the way we do business Capturing of best practices
2008–2014	White and blue collar	R&D maintained	Reduced or eliminated	Long term	Use of competing constraints Recognize importance of business value Heavy use of risk management Portfolio project management Better project governance Exit champions

2.2 EXECUTIVE VIEW OF PROJECT MANAGEMENT

The companies previously are performing strategic planning for project management and are focusing heavily on the future. They have accepted most of the principles of PM 2.0. Years ago, senior management provided lip service to project management. Today, senior management maintains a different view of project management, as seen in the Table 2-2.

TABLE 2-2 Executive View of Project Management

PM 1.0	PM 2.0
Project management should be considered as a career path position	Project management is a strategic or core competency necessary for the survival of the firm
We need people certified in project management	We need people certified in both project management and business processes
Project managers are used for project execution or delivery only	Project managers should be involved in project identification, project selection, portfolio management, and capacity-planning activities
Strategy and execution are separate activities	Project managers are the people who bridge strategy and execution

Perhaps the biggest reason why executives have accepted PM 2.0 is that they now realize that project management is a business process rather than just a project management process. As a project manager, you are no longer being paid to produce and manage deliverables. Instead, you are managing part of a business and therefore you are expected to make both project-based decisions and business decisions.

When discussing the benefits of project management for their firms, words that are commonly used by executives as a result of the growth of PM 2.0 include:[1]

- Creating a multicultural, global approach
- Better governance
- Continuous improvement opportunities
- Credibility as a trusted partner
- Optimize delivery efficiency
- Repeatable success
- Create a more integrated, agile organization
- Simplify and automate
- Make fact-based decisions

Project management is no longer regarded as a part-time occupation or even a career path position. It is now viewed as a strategic competency needed for the survival of the firm. Superior project management capability can make the difference between winning and losing a contract. To illustrate how important project management is to customers, consider the following four requirements that now appear in many requests for proposals (RFPs):

- Show us the number of PMP®s in your company and identify which PMP® will manage this contract if you are the winner through competitive bidding.
- Show us that you have an enterprise project management methodology that has a history of providing repeated successes.
- Show us the maturity level of project management in your company and identify which project management maturity model you used to perform the assessment.
- Show us that you have a best practices library for project management and that you are willing to share this knowledge with us as well as the best practices you discover on our project.

For more than 20 years, becoming a PMP® was seen as the light at the end of the tunnel. Today, that has changed. Becoming a PMP® is the light at the entryway to the tunnel. The light at the end of the tunnel requires multiple certifications. As an example, after becoming a PMP®, a project manager may desire to become certified in:

- Program management
- Business analyst skills or business management
- Business processes
- Managing complex projects
- Six Sigma
- Risk management

---

[1] Adapted from H. Kerzner, *Project Management Best Practices; Achieving Global Excellence*, 3rd ed., Wiley and International Institute for Learning, New York, 2014, pp. 13–17.

Companies such as IBM have certification boards which meet frequently and discuss which certification programs would be added value for their project managers. Certification programs which require specific knowledge of company processes or company intellectual property may be internally developed and taught by the company's own employees.

Executives have come to the realization that there is a return on investment on project management education. As such, executives are now investing heavily in customized project management training, especially in the behavioral courses. As an example, one executive commented that he felt that presentation skills training was the highest priority for his project managers. If a project manager makes a highly polished presentation before the client, the client believes that the project is being managed the same way. If the client makes a poor presentation, then the client might believe the project is managed the same way. Other training programs that executives feel would be beneficial as part of PM 2.0 for the future include:

- Managing complex projects
- Establishing KPIs and dashboard displays
- How to perform feasibility studies and cost–benefit analyses
- How to validate and revalidate project assumptions
- How to establish project governance
- How to manage multiple stakeholders
- How to design and implement "fluid" or adaptive enterprise project management methodologies
- How to develop coping skills and stress management skills

2.3 ENGAGEMENT PROJECT MANAGEMENT

With project management viewed as a strategic competency as it is now with PM 2.0, it is natural for these companies to be strong believers in "engagement project management" or "engagement selling." Years ago, the sales force would sell a product or service to a client and then move on to find another client. Today, the emphasis is on staying with the client and looking for additional work from the same client.

In a marital context, an engagement can be viewed as the beginning of a life-long partnership. The same holds true with engagement project management. Companies like IBM and HP no longer view themselves as selling products or services. Instead, they view themselves as business solution providers for their clients, and you cannot remain in business as a business solution provider without having superior project management capability.

As part of engagement project management, you must convince the client that you have the project management capability to provide solutions to their business needs on a repetitive basis. In exchange for this, you want the client to treat you as a strategic partner rather than as just another contractor.

Decades ago, the sales force (and marketing) had very little knowledge about project management. The role of the sales force was to win contracts, regardless of the concessions that had to be made. The project manager then "inherited" a project which may have included an underfunded budget and an impossible schedule. Today, sales and marketing must understand project management and be able to sell it to the client as part of engagement selling. The sales force must sell the company's project management methodology and the accompanying best practices. Sales and marketing are now part of and integrated with project management in PM 2.0.

Engagement project management benefits both the buyer and the seller, as seen in Table 2-3.

The benefits of engagement project management are clear:

- Both the buyer and the seller save on significant procurement costs by dealing with single-source or sole-source contracts without having to go through a formalized bidding process.
- Because of the potential long-term strategic partnership relationship, the seller is interested in the lifetime value of the business solution rather than just the value at the end of the project.
- You can provide life-long support to your clients as they try to develop value-driven relationships with their clients.
- The buyer will get access to many of the project management tools used by the seller. The corollary is also true.

Some companies today are establishing an engagement manager to work alongside the project manager. The engagement manager focuses more on placating the customer than on execution of the project. The relationship between the engagement manager and the project manager is shown in Table 2-4.

TABLE 2-3 Before- and After-Engagement Project Management

PM 1.0 (Before Engagement)	PM 2.0 (After Engagement)
Continuous competitive bidding	Sole-source or single-source contracting resulting in fewer suppliers to deal with
Focus on near-term value of the deliverable	Focus on lifetime value of the deliverable
Contractor provides minimal support for client's customers	Contractor supports client and client's customers through customer value analysis (CVA) and customer value measurements (CVMs)
Utilization of one often inflexible and linear enterprise resource planning system	Access to contractor's many nonlinear systems

TABLE 2-4 **Comparison of Engagement and Project Management Positions**

Customer Value Management Activities	Engagement Manager	Project Manager
Phase 1: Identifying the right customers	Strategic marketing Proposal preparation Engagement selling	Assist in proposal preparation May report to the engagement manager
Phase 2: Developing the right relationships	Define acceptance criteria (metrics/KPIs) Risk mitigation planning Client briefings Client invoicing Client satisfaction feedback and customer relations management	Supporting customer relations management Establishing performance metrics Measuring customer value and satisfaction Improving customer satisfaction management
Phase 3: Maintaining retention	Conducting customer satisfaction management meeting Updating client metrics and KPIs	Attending customer satisfaction management meetings Looking for future areas of improvement

2.4 GROWTH OF MORE COMPLEX PROJECTS

After more than three decades of using PM 1.0, we have become experts in how to manage traditional projects. These traditional projects can be for internal as well as external clients. With these projects, the statement of work is reasonably well defined, the budget and schedule are realistic, reasonable estimating techniques were used, and the final target of the project is stationary. We use a project management methodology that has been developed and undergone continuous improvements after use on several projects. This traditional methodology focuses on linear thinking; we follow the well-defined life-cycle phases and we have forms, templates, checklists, and guidelines for each phase.

Now that we have become good at these traditional projects, we are focusing our attention on the nontraditional or complex projects. One of the reasons for the growth of PM 2.0 has been the necessity to work on complex projects. However, the reader must understand that even traditional projects can be viewed as being complex under the right circumstances. Table 2-5 shows some of the differences between managing traditional and nontraditional or more complex projects.

Companies like IBM, HP, Microsoft, and Siemens are mastering complex projects, investing heavily to become solution providers, and assisting clients on a worldwide basis on managing nontraditional, complex projects. Some other distinguishing characteristics of complex projects are:

- Working with a large number of stakeholders and partners, all at different levels of project management maturity, and many of whom may not even understand the technology of the project
- Dealing with multiple virtual teams located across the world and where decisions on the project may be made in favor of politics, culture, or religious beliefs

TABLE 2-5 Comparison of Traditional and Complex Projects	
Traditional Projects (PM 1.0)	**Complex Projects (PM 2.0)**
Single-person sponsorship	Governance by committee
Possibly a single stakeholder	Multiple stakeholders
Project decision making	Project and business decision making
Inflexible project management methodology	Flexible or "fluid" project management methodology
Well-defined statement of work	Constantly evolving statement of work
Stationary target	Moving target
Periodic status reporting	Real-time status reporting
Success defined by triple constraints	Success defined by competing constraints and business value
KPIs derived from the EVMS	Unique, value-driven KPIs

- Long-term projects that begin with an ill-defined scope and undergo numerous scope changes and where the end point is a moving rather than a stationary target
- Working with partners and stakeholders who may have limited project management tools and antiquated processes that are incompatible with the project manager's tool kit

2.5 NEED FOR ADDITIONAL METRICS

Birth of metric-driven project management.
Copyright © Scott Maxwell/Fotolia

For almost five decades, project decisions revolved around only two metrics: time and cost. The result was that decision-makers were making seat-of-the-pants decisions rather than informed decisions. Everybody knew that more metrics were needed, but measurement techniques for the additional metrics were just in the infancy stages. Fact-based decision making was seen as a luxury rather than the norm.

Project managers in the future will have an abundance of metrics from which to select. We will have metric libraries. The libraries will be managed by the PMO. Employees will be trained in how to identify, track, quantify, and report metrics. Many of the metrics will be identified by the client and the stakeholders as requirements for them to conduct their informed decision-making activities.

The PMO will have the responsibility for continuous improvements on the metrics the same way it is performed on the best practices that are captured. Because the project managers will be expected to make business as well as project decisions, project managers will be expected to work with both project and business metrics.

With PM 2.0, we are heading into an era of metric-driven project management. Project plans and baselines will be established around the metrics selected for a project. Each project can have a unique set of project metrics. The metrics may be valid for just one or two life-cycle phases rather than the entire project. There may also be established metrics to be used after the project is over or "goes live."

2.6 NEW DEVELOPMENTS IN PROJECT MANAGEMENT

For companies to be successful at managing complex projects on a repetitive basis and function as a solution provider, the project management methodology and accompanying tools must be fluid or adaptive. This is a significant change from PM 1.0. This means that you may need to develop a different project management methodology or apply your existing methodology differently to interface with each stakeholder given that each stakeholder may have different requirements and expectations and most complex projects have long time spans. Figure 2-1 illustrates some of the new developments in project management: The five items in the figure fit together when done properly:

New Success Criteria: At the initiation of the project, the project manager will meet with the client and the stakeholders to come to a stakeholder agreement on what constitutes success on this project. Initially, many of the stakeholders can have their own definition of success, but the project manager must forge an agreement.

Metrics and Key Performance Indicators: Once the success criteria are agreed upon, the project manager will work with the stakeholders to define the KPIs that each stakeholder wishes to track. It is possible that each stakeholder will have different metric/ KPI requirements.

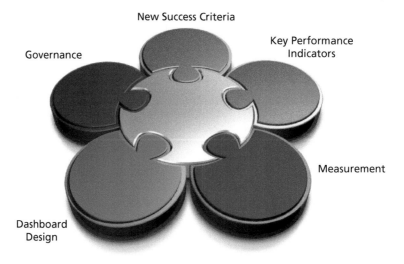

FIGURE 2-1 **Future trends in project management**

Measurement: Updating the dashboards and the metrics/KPIs requires measurement. This is the hardest part because not all of the team members or strategic partners may have the capability to track all of the metrics/KPIs.

Dashboard Design: Once the metrics/KPIs are identified and measured, the project manager, along with the appropriate project team members and a dashboard designer, will prepare a dashboard for each stakeholder. Some of the metrics/KPIs in the dashboards will be updated periodically whereas others may be updated on a real-time basis.

Governance: Once the measurements are made, the information is provided to the governance committee for any decisions necessary. The governance board can include key stakeholders as well as stakeholders who are just observers.

2.7 PROJECT MANAGER'S TOOL BOX

What tools are in the project manager's toolbox for PM 2.0?

With PM 1.0, project managers had limited tools. Other than an EVMS to accompany the project management methodology, many project managers relied upon their own ingenuity. With PM 2.0, project managers can have as many as 50 tools with which to work. Some of the tools are:

- Earned value measurement system
- Corporate project management policies and procedures
- Flexible templates (for all domain areas)
- Flexible forms (for all domain areas)
- Flexible guidelines (for all domain areas)
- Flexible checklists (for all domain areas)
- Flexible flowcharts (for all domain areas)
- Playbooks for each type of project
- Playbooks for particular clients
- Advanced project and business metrics
- Advanced project and business KPIs
- Use of a best practices library
- Use of a metrics/KPI library
- Use of a knowledge management library
- Network diagrams
- Historical data for estimating
- Industry-specific databases
- Behavioral software
- Benefit-to-cost templates
- Early warning indicators
- Crisis dashboards
- Risk management templates
- Failure analysis templates
- Assessment instruments

2.8 NEED FOR CONTINUOUS IMPROVEMENT

Who will be responsible for continuous improvements with PM 2.0?

Copyright © Scott Maxwell/Fotolia.

With PM 1.0, continuous improvement activities, if done at all, were accomplished by the project managers. Lessons learned and best practices were captured from successes only because people were afraid to openly admit mistakes. Unfortunately, we tend to capture significantly more best practices from failures than from successes. The PMO may have been given added responsibility to search for the best practices in addition to maintaining the best practices library. It was also common practice to have professional facilitators assigned to the PMO solely for the purpose of debriefing teams at the end of the project for the capturing of best practices.

With PM 2.0 and the execution of much more complex and larger projects, it is impossible to wait until the projects are completed to capture the best practices. Everyone on the project team must take the responsibility to seek out lessons learned and best practices throughout the project. Waiting until the project is over or just looking at best practices at some of the gate reviews, which could be a year apart in length, might delay the benefits of the best practices from being immediately implemented.

Today, with the abundance of metrics available to project team members, the opportunities for discovery of best practices are endless. Team members will attend training courses in metrics management and part of each course may be devoted to ways that metrics can be used to discover best practices.

2.9 CONCLUSIONS

The future of project management may very well rest in the hands of the solution providers. These providers will custom design project management methodologies for each client and possibly for each stakeholder. They must be able to develop project management skills that go well beyond the current *PMBOK® Guide* and demonstrate a willingness to make business decisions as well as project decisions. The future of project management looks quite good, but it will be a challenge.

DISCUSSION QUESTIONS

The discussion questions are for classroom use to stimulate group thinking about PM 2.0. There are no right or wrong answers to most of the questions.

1. Why do recessions promote advances in project management?
2. What are the primary factors that encourage senior management to accept the risks associated with the undertaking of more complex projects?
3. What do you believe is the single most important factor that changed the executive's view of project management?
4. Why has the need for continuous improvements in project management grown so rapidly?
5. Are the tools in a project manager's toolbox generic or industry specific?
6. Why has the need for additional metrics grown so fast?

7. Other than a recession, what has caused significant changes in project management?
8. Who defines whether or not a project is complex?
9. Can the project manager position and engagement manager position be merged into one position? What are the pros and cons?
10. Can the need for project infographic designers be eliminated in PM 2.0?

CHAPTER 3

UNDERSTANDING SUCCESS AND FAILURE

3.0 INTRODUCTION

One of the most significant differences between PM 1.0 and PM 2.0 has been the way that we define project success and project failure. With PM 1.0, it was quite common for both the customer and the contractor to be working toward different definitions of success and failure. The contractor's definition of success was profitability and customer's definition of success was an acceptable deliverable.

Project management evolved from a set of processes that were once considered "nice" to have to a structured methodology that is today considered mandatory for the survival of the firm. Companies are now realizing that their entire business, including most of the routine activities, can be regarded as a series of projects. Simply stated, we are managing our business by projects and the need for PM 2.0 is quite apparent. As such, the definitions of success and failure have changed as have the metrics used to define success and failure.

Today, as the relative importance of project management permeates each facet of the business, knowledge is captured on best practices in project management. This includes better ways to define project success and failure. Some companies view this knowledge as intellectual property to be closely guarded in the vaults of the company. Others share this knowledge in hope of discovering other best practices that improve the chances for project success. Companies are now performing strategic planning for project management. Project management is viewed as a core competency for the firm.

One of the benefits of performing strategic planning for project management is that it usually identifies the need for capturing and retaining best practices. This includes best practices related to using specific metrics. Unfortunately this is easier said than done. One of the reasons for this difficulty is that companies today are not in agreement on the definition of a best practice, nor do they understand that best practices lead to continuous improvement, which in turn leads to the capturing of more best practices. The same can be said for the difficulty in getting all parties to agree on the definition of success and failure.

3.1 PROJECT MANAGEMENT—EARLY YEARS: 1945–1960

During the 1940s, line managers wore multiple hats and functioned as part-time project managers using the concept of over-the-fence management to manage projects. Each line manager, wearing the secondary hat of a project manager, would perform the work necessitated by their line organization and, when completed, would throw the "ball" over the fence in hopes that someone would catch it in the downstream part of the chain. Once the ball was thrown over the fence, the line managers would wash their hands of any responsibility for the project because the ball was no longer in their yard. If a project failed, blame was placed on whichever line manager had the ball at that time. Everyone in the chain had their own definition of success and failure. Success could be defined simply as creating technology, maintaining a budget and schedule regardless of the quality of the deliverable, or just completing an assignment and moving on to another project.

The problem with over-the-fence management was that the customer had no single contact point for questions. The filtering of information wasted precious time for both the customer and the contractor. Customers who wanted firsthand information had to seek out the person in possession of the ball. Time and cost were the only two metrics being considered and reported. For small projects, this was easy and accepted as a way of life. But as projects grew in size and complexity, this became more difficult. There was no consensus on the definition of success and failure.

During this time period, which was the beginning of PM 1.0, very few project management best practices were identified. If there were best practices, then they would stay within a given functional area, never to be shared with the remainder of the company. Each functional department may have had their own tracking metrics. Suboptimal project management decision making was the norm, as was the definition of success.

Following World War II, the United States entered into the Cold War. To win a Cold War, one must compete in the arms race and rapidly build weapons of mass destruction that could act as a deterrent force. The victor in a Cold War was the one who could retaliate with such force as to obliterate the enemy. Developing weapons of mass destruction was a very large project involving potentially thousands of contractors. The definition of success was the creation of the technology, not budgets or schedules.

The arms race made it clear that the traditional use of over-the-fence management would not be acceptable to the Department of Defense (DOD) for projects such as the B52 bomber, the Minuteman intercontinental ballistic missile, and the Polaris submarine. The government wanted a single point of contact within each contractor's organization, namely, a project manager who had total accountability throughout all project phases. In addition, the government wanted the project manager to possess a command of technology rather than just an understanding of technology, which mandated that the project manager be an engineer, preferably with an advanced degree in some branch of technology. The use of project management was then mandated for some of the smaller weapon systems such as jet fighters and tanks. NASA mandated the use of project management for all activities related to the space program.

The DOD tried unsuccessfully to define project success as completion within time and cost. But some projects in the aerospace and defense industries were having cost overruns in excess of 200–300%. Blame was erroneously placed upon improper implementation of project management when, in fact, the real problem was the inability to forecast technology, resulting in numerous scope changes occurring. Forecasting technology is extremely difficult for projects that could last 10–20 years. With engineers placed in charge of projects, project success was identified solely in technical terms. Simply stated, did it work? Maintaining cost and schedule was nice to have but not at the expense of technology. And as long as the DOD was willing to pay for the cost overruns and let the project schedules slip, there was no pressure placed upon organizations to consider any definition of success other than technical accomplishment.

By the mid- to late 1960s, the aerospace and defense industries were using project management on virtually all projects, and they were pressuring their suppliers to use it as well. Project management was growing, but at a relatively slow rate except for aerospace and defense.

Because of the vast number of contractors and subcontractors, the government needed standardization for project planning, execution, monitoring and control, and reporting of information. The government established a life-cycle planning and control model and a cost-monitoring system, later to be called the earned value measurement system (EVMS), and created an organization of project management auditors to make sure that the government's money was being spent as planned. These practices were to be used on all government programs above a certain dollar value. Private industry viewed these practices as an over-management cost and saw no practical value in project management if these were the established processes. There were also misconceptions concerning project management. Some of the misconceptions were:

- Project management is merely a scheduling tool, such as program evaluation and review technique/critical path method (PERT/CPM) scheduling.
- Project management applies to large projects only.
- Project management is designed for large government projects only.
- Project managers must be engineers and preferably with advanced degrees.
- Project managers need a "command of technology" to be successful.
- Project success is measured in technical terms only (i.e., did it work?).
- Only time and cost metrics may be needed.
- Slippages on time and cost are acceptable.

3.2 PROJECT MANAGEMENT BEGINS TO GROW: 1970–1985

During this time period, with a better understanding of project management, the growth of project management had come about more through necessity than through desire, but at a very slow rate. Its slow growth was attributed mainly to lack of acceptance of the new management techniques necessary for its successful implantation. An inherent fear of the unknown acted as a deterrent for both managers and executives.

Other than aerospace, defense, and construction, the majority of the companies in the 1960s supported a more informal approach for managing projects. With informal project management, just as the words imply, the projects were handled on an informal basis whereby the authority of the project manager was minimized. This made it difficult for the project managers to make the necessary decisions to conform to project success criteria. Most projects were handled by functional managers and stayed in one or two functional lines, and formal communications channels were deemed either unnecessary or handled informally because of the good working relationships between line managers. Those individuals that were assigned as project managers soon found that they were functioning more as project leaders or project monitors than as real project managers. For the employees assigned to these projects, the definition of success was defined as satisfying the needs of the functional managers for fear that negative results could impact one's performance reviews. Each functional manager had their own definition of success and quite often the functional manager's definition of success was contradictory to the project manager's definition of success. This was a prime characteristic of PM 1.0.

By 1970 and through the early 1980s, more companies departed from informal project management and restructured to formalize the project management process, mainly because the size and complexity of their activities had grown to a point where they were unmanageable within the current structure. Standardized project management processes were needed, including a better way of defining success.

3.3 GROWTH IN COMPETING CONSTRAINTS

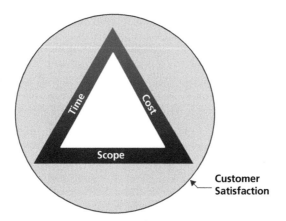

Historically, the definition of success had been meeting the customer's expectations of time and cost regardless of whether or not the customer was internal or external. This was a characteristic of PM 1.0. Success was defined as getting the job done within the constraints of time, cost, and scope according to the triple constraints shown in Figure 3-1. In some industries, completing projects within time, cost, and scope was almost impossible. Trade-offs were needed and the definition of success then included customer acceptance of the project, as shown by the circle in Figure 3-1. In some companies, customer acceptance was considered as the only constraint that was important even though other constraints existed. With PM 2.0, we readily admit that many constraints are possible, but we prioritize the importance of each constraint. The importance of each constraint can vary from project to project.

Very few projects are ever completed without trade-offs or scope changes on time, cost, and scope. Therefore, success could still occur without exactly hitting this singular

How do we determine the number of constraints on the projects?

Copyright © Scott Maxwell/LuMaxArt/Shutterstock

Time

Cost

Scope

Customer Satisfaction

FIGURE 3-1   Triple constraints

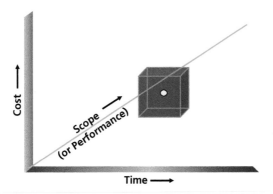

FIGURE 3-2   Boundary box for success

point. In this regard, success could be defined as a cube, such as seen in Figure 3-2. The singular point of time, cost, and scope (or quality) would be a point within the cube or boundary box, constituting the convergence of the critical constraints for the project.

As long as the project was completed within the boundary box of constraints, the project could be regarded as a success. You could be two weeks late and the customer would still accept the project's results. You could be over budget by 10% and still treat the project as a success. The customer might also view the project as a success if only 95% of the specifications were met.

As project management grew and people recognized the need for PM 2.0 practices, companies realized that there were more than three constraints on most projects and that the definition of success may need to consider meeting all of the constraints. The constraints were then identified as primary and secondary constraints, as shown in Table 3-1.

The hard part was identifying the number of constraints and their relative importance. Project managers assumed that the project's business case would identify the number of constraints, but this did not happen often.

TABLE 3-1 Primary and Secondary Constraints

Primary Constraints	Secondary Constraints
Within time	Winning follow-on work from this client
Within cost	Using the client's name as a future reference
Within scope or quality	Commercialization of a product
Accepted by the customer	Within minimum or mutually agreed-upon scope changes
	Maintaining the company's image and reputation
	Without disturbing the main flow of daily work
	Without changing the corporate culture
	Without violating the company's social responsibility
	Maintaining client goodwill

(continued)

TABLE 3-1 (*Continued*)

Primary Constraints	Secondary Constraints
	Creating business value
	Without violating safety requirements
	Providing efficiency and effectiveness of operations
	Maintaining ethical conduct
	Satisfying OSHA and EPA requirements
	Providing strategic alignment
	Maintaining regulatory agency relations
	Effective use of risk mitigation strategies

3.4 RULE OF INVERSION

Is it easy to determine the true status of the project?

Copyright © Scott Maxwell/Fotolia.

The secondary constraints posed a problem. When the EVMS was first created, everyone knew that time and cost alone could not and should not be used as the only success criteria. Describing the status of the project using just time and cost was like rolling the dice; there's a chance that you may be right but there is a greater chance that you may be wrong. Using only time and cost metrics implies that all performance measurements are linear. Percent complete is not linear with time and cost measurements. Everyone knew that other metrics should be used. Unfortunately, the rule of inversion was adopted, which stated that the metrics used to define success should be the easiest metrics to track and report. Time and cost were of course the easiest metrics to track and report. Measurement techniques had not matured to the point where the secondary constraints could be accurately measured and tracked. Table 3-2 shows how we viewed the complexity of several metrics that can possibly assist in defining project success.

TABLE 3-2 **Metric Measurement Complexity**

Metric	Measurement Complexity
Profitability	Easy
Customer satisfaction	Hard
Goodwill	Hard
Penetrate new markets	Easy
Develop new technology	Medium
Technology transfer	Medium
Reputation/Image	Hard
Stabilize work force	Easy
Efficiency, effectiveness, and productivity	Hard
Utilize unused capacity	Easy

3.5 GROWTH IN MEASUREMENT TECHNIQUES

The growth in measurement techniques, which led in part to the development of PM 2.0, made it somewhat easier to identify the correct number of constraints and perform tracking. Doug Hubbard states four useful assumptions when considering measurement techniques[1]:

- It has been measured before.
- You have more data than you think.
- You need less data than you think.
- Useful new observations are more accessible than you think.

What is the best way to measure performance against constraints?

Copyright © Scott Maxwell / Fotolia.

Advances in measurement techniques have given us several options for measuring any metric. Typical measurement methods include:

- Observations
- Ordinal (i.e., four or five stars) and nominal (i.e., male or female) data tables
- Ranges/sets of value
- Simulation
- Statistics
- Calibration estimates and confidence limits
- Decision models [e.g, expected value (EV), expected value of perfect information (EVPI)]
- Sampling techniques
- Decomposition techniques
- Human judgment
- Rules (e.g., 50/50, 80/20, 0/100, % complete)

Not all project goals or constraints are tangible. Intangible goals and constraints may be tough to measure, but they are not immeasurable. Tough things to measure include:

- Collaboration
- Commitment
- Creativity
- Culture
- Customer satisfaction
- Emotional maturity
- Employee morale
- Image/reputation
- Leadership effectiveness
- Motivation
- Quality of life
- Stress level
- Sustainability
- Teamwork

---

[1] D. W. Hubbard, *How to Measure Anything; Finding the Value of Intangibles in Business*, 3rd ed., Wiley, Hoboken, NJ, 2014, p. 59

In an ideal situation, all measurements would be quantitative measurements. Unfortunately, many items in the above list are measurable, but qualitatively. Some people argue that qualitative measurements are just a subset of quantitative measurements. As we get better at metric measurement, perhaps all measurements will eventually be quantitative.

3.6 TRADE-OFFS

Which trade-offs should I make?

Copyright © Scott Maxwell/
LuMaxArt/Shutterstock.

When managing a project according to the triple constraints of time, cost, and scope, we perform a juggling act and often find a way to meet all three constraints, all of which usually carry an equal degree of importance. When the number of constraints increases to five or six constraints, it may be difficult to meet all of the constraints and a prioritization of constraints may be necessary.

The prioritization of constraints can change over the life of the project based upon the needs of the project manager, the client, and the stakeholders. Changing the priorities of the constraints can lead to scope changes and play havoc with the requirements and baselines. There must be a valid reason for changing the prioritization of the constraints after project go-ahead.

Although many projects are completed successfully, at least in the eyes of the stakeholders, the final criteria from which success is measured may be different than the initial criteria because of trade-offs. It may not be possible to complete the project within the original success criteria of time, cost, and scope. Today, we realize that there can be multiple constraints on a project and, rather than using the terminology of the triple constraints, we focus our attention on competing constraints. This makes it even more complicated when trade-offs must be performed.

It is a mistake to believe that, especially on long-term projects, the final success criteria will be the same as the initial success criteria. Changes can occur. These changes can occur any time during the life of a project and can then cause trade-offs in the competing constraints, thus requiring that changes be made to the success criteria. In an ideal situation, we would perform trade-offs on any or all of the competing constraints such that acceptable success criteria would still be met.

As an example, assume that a project was initiated using the success criteria of the triple constraints shown on the sides of the triangle on the left in Figure 3-3. Part way through the project, the enterprise environmental factors change, a new senior management team is brought in with their own agenda, or a corporate crisis occurs such that the credibility of the corporation is at stake. In such a case, additional competing constraints of image/reputation, quality, risk, and value are introduced. With PM 1.0, these new constraints were inserted into the center of the triangle as shown in Figure 3-3 and the project manager simply determined the impact that these additional constraints would have on the three sides of the triangle. Only the three constraints on the sides of the triangle were tracked and used to define success.

With PM 2.0, all of the constraints are tracks and even prioritized. As such, after prioritization, the sides of the triangle on the right in Figure 3-3 can be more important than the original triple constraints. However, there can be significantly more than three competing constraints that are a high priority in which some geometric shape other than a triangle might work best.

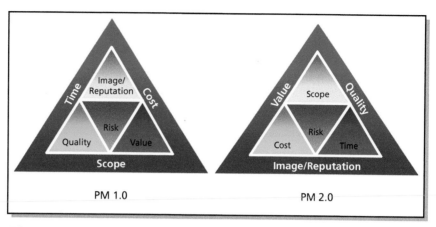

FIGURE 3-3 **Competing constraints**

Prioritizing constraints is a necessity, especially if more than three constraints exist. For example, years ago, in Disneyland and Disneyworld, the project managers designing and building the attractions at the theme parks had six constraints:

- Time
- Cost
- Scope
- Safety
- Aesthetic value
- Quality

At Disney, the last three constraints of safety, aesthetic value, and quality were considered locked-in constraints that could not be altered during trade-offs. All trade-offs were made on time, cost, and scope. Some constraints simply cannot change while others may have flexibility.

Not all constraints are equal in importance throughout the life of a project. For example, in the initiation phase of a project, scope may be the critical factor and all trade-offs are made on time and cost. During the execution phase of the project, time and cost may become more important and then trade-offs will be made on scope.

3.7 PUTTING TOGETHER COMPONENTS OF SUCCESS

We mentioned previously that in PM 2.0 each project may have a different definition of success and the importance of the customer and the contractor to come to an agreement on the success definition. Metrics must be established for each of the success criteria.

Assume that there are eight constraints on a project and all eight constraints make up the success criteria. Will the project still be regarded as a success if:

- Seven of the eight success criteria are met but the project is completed late by two weeks?
- Seven of the eight criteria are met but there is a cost overrun of $50,000?
- The client asked for 10 deliverables per month and we shipped 8 deliverables per month, and the rest were shipped late?

How many puzzle pieces are needed to define success?

The boundary box described in Figure 3-2 showed that success can be regarded as a geometric shape or boundary box, and as long as the constraint metrics remain in the boundary box, success is achieved.

But what happens if not all of the competing constraints can be met? As the number of constraints increases, so does the likelihood that not all constraints can be met. Loosening the constraints may be an option but can lead to an improper evaluation of ongoing success.

With PM 2.0, success may be viewed as putting together pieces of a puzzle. It may not be necessary to have all of the pieces of the puzzle in place to see what the picture looks like. The same holds true for defining success. While we would prefer to have all of the constraints met, we may very well define success as meeting just a percentage of the constraints or we may say that only five of the eight constraints make up the success criteria.

The success criteria may be made up of primary and secondary constraints such as was defined in Table 3-1. The agreed-upon success criteria may be a composition of the primary constraints whereas the secondary constraints and their importance may be unique to the customer or the contractor. As an example, both the customer and contractor may be in agreement that cost containment is a primary constraint and part of the success criteria. But the contractor may have a secondary constraint of maintaining profit margins and this constraint is not part of the success criteria. Likewise, the customer may have a secondary constraint of how much value this project brings to their clients, but this may not be an agreed-upon component of the success criteria.

3.8 NEW DEFINITION OF SUCCESS

How was the definition of success changed?

For decades, we have used the same definition for a project and for the success of a project. There were some minor changes to the definitions over the years, but the changes were small. Today, value is an important element in both definitions.

Definition of Project

- **PMBOK® Guide Definition:** A temporary endeavor undertaken to create a unique product, service, or result (**PMBOK® Guide,** Fifth Edition, Glossary).
- **Future Definition:** A collection of sustainable business value scheduled for realization.

Definition of Project Success

- **Traditional Definition:** Completion of the project within the triple constraints of time, cost, and scope.
- **Future Definition:** Achieving the desired business value within the competing constraints.

For years, value was considered an elusive term and very little effort was expended in the measurement of value. Today, value can be measure and reported on a project status dashboard.

In Figure 2-1 and the previous section, we showed that, in the future, customers and contractors must work together to come up with an agreed-upon definition of project success. It is entirely possible that each project can have a different success criterion and a different interpretation of value, thus requiring possibly a different set of metrics to confirm that success has been achieved.

Business value may become the driver in defining success. As stated previously, the more constraints you have, the greater the likelihood that not all of the constraints can be met. Therefore, if business value creation is the primary driver for success, then all or part of the competing constraints may be of secondary importance.

3.9 UNDERSTANDING PROJECT FAILURE[2]

Most companies seem to have a relatively poor understanding of what is meant by project failure. Project failure is not necessarily the opposite of project success. Simply because we could not meet the project's success criteria or adhere to all of the competing constraints is not an indication that the project was a total failure. Consider the following example.

Situation: During an internal meeting to discuss the health of various projects undertaken to create new products, a vice president complained that less than 20% of the R&D projects were successful and reached the product commercialization stage. He then blamed poor project management for the failures of the other 80% of the projects. The director of the project management office then spoke up asserting that most of the other 80% of the projects were not failures. They had in fact created intellectual property that was later used on other R&D projects (i.e., spinoffs) to create commercially successful products.

Not all projects can be saved

Copyright © Scott Maxwell/ LuMaxArt/Shutterstock.

The above example should make it clear that the definition of project failure is more of a grey area than pure black and white. If knowledge and/or intellectual property is gained on the project, then perhaps the project should not be considered as a complete failure. All project managers know that things may not always go according to plan. Replanning is a necessity in project management. We can begin a project with the best of intentions and prepare a plan based upon the least risk. Unfortunately, the least risk plan usually requires more time and more money. If the project must be replanned using least time as the primary success criterion, then we must be willing to incur more risk and perhaps additional costs.

There is no universally accepted diagnosis as to why projects fail because each project has its own set of requirements, its own unique project team, and its own success criteria and can succumb to changes in the enterprise environmental factors. Failures can and will happen on some projects regardless of the company's maturity level in project management. As seen in Figure 3-4, it often takes companies two years or longer to become reasonably good at project management and perhaps another five years to reach some degree of excellence. Excellence in project management is defined as a continuous stream of projects that meet the company's project success criteria.

---

2 Adapted from H. Kerzner, *Project Recovery: Case Studies and Techniques for Overcoming Project Failure*, Wiley and International Institute for Learning, New York, 2014, pp.12–16

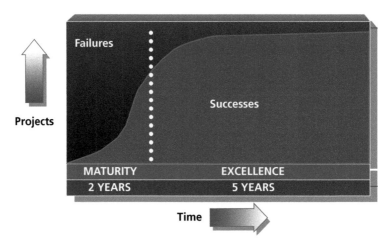

FIGURE 3-4 Some projects will fail

However, as seen in Figure 3-4, even with a high degree of project management excellence some projects can and will fail. There are three reasons for this:

- Any executive who always makes the right decision certainly is not making enough decisions.
- Effective project management practices can increase your chances of project success but cannot guarantee that success will be achieved.
- Business survival is often based upon how well the company is able to accept and manage business risks. Knowing which risks are worth accepting is a difficult decision.

One of the most commonly read reports on why IT projects fail is the Chaos Report prepared by the Standish Group. The Chaos Report identifies three types of IT project outcomes:

- **Success:** A project that gets accolades and corporatewide recognition for having been completed on time and within budget and meeting all specification requirements.
- **Challenged:** A project that finally reaches conclusion, but there were cost overruns and schedule slippages, and perhaps not all of the specifications were met.
- **Failure:** A project that was abandoned or canceled due to some form of project management failure.

It is interesting to note how quickly IT personnel blame project management as the primary reason for an IT failure. Although these categories may be acceptable for IT projects, it may be better to use the following breakdown for all projects in general:

- **Complete Success:** The project met the success criteria, value was created, and all constraints were adhered to.
- **Partial Success:** The project met the success criteria, the client accepted the deliverables, and value was created, although one or more of the success constraints were not met.

- **Partial Failure:** The project was not completed as expected and may have been canceled early on in the life cycle. However, knowledge and/or intellectual property was created that may be used on future projects.
- **Complete Failure:** The project was abandoned and nothing was learned from the project.

The following situations provide examples of each of these categories.

Situation: A company undertook a 1-year R&D project designed to create a new product. Assuming the product could be developed, the company had hoped to sell 500,000 units over a 2-year period. During the R&D effort, the R&D project team informed management that they could add significant value to product if they were given more money and if the schedule were allowed to slip by about 6 months. Management agreed to the schedule slippage and the cost overrun despite resistance from sales and marketing. More than 700,000 units were sold over the first 12 months after product release. The increase in sales more than made up for the cost overrun.

Lessons learned: In this situation, the project was considered a complete success even though there was a schedule slippage and a cost overrun. Significant value was added to the business.

Situation: A company won a contract through competitive bidding. The contract stipulated that the final product had to perform within a certain range dictated by the product's specifications. Although there were no cost overruns or schedule slippages, the final product could meet only 90% of the specification's performance requirements. The client reluctantly accepted the product and later gave the contractor a follow-on contract to see if they could reach 100% of the specification's performance requirements.

Lessons learned: This situation was considered a partial success. Had the client not accepted the deliverable, then the project may have been classified as a failure.

Situation: A company had a desperate need for software for part of its business. A project was established to determine whether the company should create the software from scratch or purchase an off-the-shelf package. The decision was made to purchase an expensive software package shortly after one of the senior managers in a software company made an excellent presentation on the benefits the company would see after purchasing and using the software as stated. After purchasing the software, the company realized that it could not get the expected benefits unless the software was custom designed to its business model. The software company refused to do any customization and reiterated that the benefits would be there if the software was used as stated. Unfortunately, it could not be used as stated, and the package was shelved.

Lessons learned: In the above situation, the company considered the project a total failure. No value was received for the money spent. Eventually the company committed funds to create its own software package customized for its business applications.

Situation: A hospital had a policy where physicians and administrators would act as sponsors on large projects even though they had virtually no knowledge about project management. Most of the sponsors also served on the committee that established the portfolio of projects. When time came to purchase software for project management applications, a project team was established to select the package to be procured. The project team was composed entirely of project sponsors who had limited knowledge of project management. Thinking that they were doing a good thing, the committee purchased a $130,000 software package with the expectation that it would be used by all of the project managers. The committee quickly discovered that the organization was reasonably immature in project management and that the software was beyond the capabilities of most project team members. The software was never used.

Lessons learned: The above situation, just like the previous situation, was considered a complete failure.

Situation: A company was having difficulty with its projects and hired a consulting company for project management assistance. The decision to hire the company was largely due to a presentation made by one of the partners who had more than 20 years of project management experience.

After the consulting contract was signed, the consulting company assigned a small team of people, most of whom were recent college graduates with virtually no project management experience. The consulting team was given offices in the client's company and use of the client's computers.

The consulting team acted merely as note-takers in meetings. The quarterly reports they provided to the client were simply a consolidation of the notes they would take during project team meetings. The consulting team was fired since they were providing no value. The client was able to recover from the company's computers several of the e-mails sent from the consultants to their superiors. One of the e-mails that came from the headquarters of the consulting company stated, "We know we didn't give you a qualified team, but do the best you can with what you have." The client never paid the consulting company the balance of the money due on the contract.

Lessons learned: In the above example, the client eventually sued the consulting company for failure to perform and collected some damages. The client considered the consulting project a complete failure.

Situation: A company worked on an R&D project for more than a year just to discover that what they wanted to do simply would not happen. However, during their research, they found some interesting results that later could be used in creating other products.

Lessons learned: Although this project was a partial failure, it did create intellectual property that could be used later.

3.10 CAUSES OF PROJECT FAILURE

What are the causes of project failure?

There are numerous causes of project failure and most failures are the result of more than one cause. Some causes directly or indirectly lead to other causes. For example, business case failure can lead to planning and execution failure. For simplicity sake, project failures can be broken down into the following categories:

Planning/Execution Failures

- Business case deterioration
- Business case requirements changed significantly over the life of the project
- Technical obsolescence has occurred
- Technologically unrealistic requirements
- Lack of a clear vision
- Plan asks for too much in too little time

- Poor estimates, especially financial
- Unclear or unrealistic expectations
- Assumptions, if they exist at all, are unrealistic
- Plans are based upon insufficient data
- No systemization of the planning process
- Planning is performed by a planning group
- Inadequate or incomplete requirements
- Lack of resources
- Assigned resources lack experience or the necessary skills
- Resources lack focus or motivation
- Staffing requirements are not fully known
- Constantly changing resources
- Poor overall project planning
- Established milestones are not measurable
- Established milestones are too far apart
- The environmental factors have changes causing outdated scope
- Missed deadlines and no recovery plan
- Budgets are exceeded and out of control
- Lack of replanning on a regular basis
- Lack of attention provided to the human and organizational aspects of the project
- Project estimates are best guesses and not based upon history or standards
- Not enough time provided for estimating
- No one knows the exact major milestone dates or due dates for reporting
- Team members working with conflicting requirements
- People are shuffled in and out of the project with little regard for the schedule
- Poor or fragmented cost control
- Weak project and stakeholder communications
- Poor assessment of risks if done at all
- Wrong type of contract
- Poor project management; team members possess a poor understanding of project management, especially virtual team members
- Technical objectives are more important than business objectives
- Assigning critically skilled workers, including the project manager, on a part-time basis
- Poor performance-tracking metrics
- Poor risk management practices
- Insufficient organizational process assets

Governance/Stakeholder Failures

- End-use stakeholders not involved throughout the project
- Minimal or no stakeholder backing; lack of ownership
- New executive team in place with different visions and goals
- Constantly changing stakeholders
- Corporate goals and/or vision not understood at the lower organizational levels
- Unclear stakeholder requirements
- Passive user stakeholder involvement after handoff

- Each stakeholder uses different organizational process assets, which may be incompatible with each other
- Weak project and stakeholder communications
- Inability of stakeholders to come to an agreement

Political Failures

- New elections resulting in a change of power
- Changes in the host country's fiscal policy, procurement policy, and labor policy
- Nationalization or unlawful seizure of project assets and/or intellectual property
- Civil unrest resulting from a coup, acts of terrorism, kidnapping, ransom, assassinations, civil war, and insurrection
- Significant inflation rate changes resulting in unfavorable monetary conversion policies
- Contractual failure such as license cancellation and payment failure

Failures can also be industry specific, such as IT failure or construction failure. Some failures can be corrected while other failures can lead to bankruptcy. If the same causes of failure keep reappearing, then metrics should be established to track these causes of failure.

DISCUSSION QUESTIONS

The discussion questions are for classroom use to stimulate group thinking about PM 2.0. There are no right or wrong answers to most of the questions.

1. What were the driving forces for the implementation of project management in the early years?
2. How do recessions impact continuous improvements in project management?
3. What factors drove the need for new definitions of project success and failure?
4. Can a project be a success and a failure at the same time?
5. Can the causes of project failure be industry specific, company specific, or project specific?
6. What caused the creation of the rule of inversion?
7. How will the new definitions of project success affect the way we do trade-offs?
8. What impact did advances in measurement techniques have on the way we now define success and failure?
9. What impact did advances in measurement techniques have on the way we perform trade-offs?
10. Who defines project success and when in the project life cycle does this definition appear?

Copyright © Scott Maxwell/LuMaxArt/Shutterstock

CHAPTER 4

VALUE-DRIVEN PROJECT MANAGEMENT

4.0 INTRODUCTION

For more than four decades, the traditional view of project management was that, if you completed the project and adhered to the competing constraints or the triple constraints of time, cost, and performance, the project was successful. Perhaps in the eyes of the project manager the project appeared to be a success. But in the eyes of the customer or even the parent company's senior management, the project might be regarded as a failure.

With PM 2.0, project managers are now becoming more business oriented. Projects are being viewed as part of a business for the purpose of providing value to both the ultimate customer and the parent corporation. Project managers are expected to understand business operations more so today than in the past. Some companies are even developing internal training programs on business processes for their project managers. And, as project managers become more business oriented, our definition of success on a project now includes a business component. The business component is directly related to value. As such, project managers must now ask themselves the following:

- Is the project creating value for the client?
- Is the project creating value for my company as well?
- Does the client recognize the value that is being created?
- How does the client measure the value that is created?

An essential success criterion with PM 2.0 is that projects must provide some degree of value when completed as well as meet the competing constraints. Perhaps the project manager's belief is that meeting the competing constraints provides value, but that is not always the case. Why should a company work on projects that provide no near- or long-term value? Too many companies either are working on the wrong projects or simply have a poor project portfolio selection process, and the value that appears at the completion of the projects is less than maximized even though the triple constraints have been met.

Assigning the best resources to projects that provide no appreciable value is an example of truly inept management and poor decision making. Yet selecting projects that will guarantee value or an acceptable return on investment (ROI) is very challenging because some of today's projects do not provide the targeted value until perhaps years into the future. This is particularly true for R&D and new product development where as many as 50 or more ideas must be explored to generate one commercially successful product. In the pharmaceutical industry, the cost of developing a new drug could run about $850 million to $1.5 billion, take 3000 days to go from exploration to commercialization, and provide no meaningful ROI. In the pharmaceutical industry, less than 3% of the R&D projects are ever viewed as a commercial success and generate more than $600 million per year in revenue.

There are many views of the definition of value, and this is quite apparent in PM 2.0. For the most part, value is like beauty; it is in the eyes of the beholder. In other words, value may be viewed as a perception at project selection and initiation based on data available at the time. But at project completion the actual value becomes a reality that may not meet the expectations that had initially been perceived.

Another problem is that the achieved value of a project may not satisfy all of the stakeholders since each stakeholder may have had a different perception of value as it relates to their business function. The definition of value can be industry specific, company specific, or even dependent on the size, nature, and business base of the firm. Some stakeholders view value as job security or profitability. Others might view value as image, reputation, or the creation of intellectual property. Satisfying all stakeholders is a formidable task often difficult to achieve and, in some cases, may simply be impossible. PM 2.0 cannot guarantee that value will exist at project completion.

When value is obtained, the company must decide how to capitalize on what has been gained. The projects and associated procedures that resulted in the creation of value can either lead to or become examples of best practices and formally advertised in organizational literature. Other forms of value may be seen as company proprietary information and intellectual property that is not to be released publicly. In any event, the ultimate goal is to achieve sustainable value.

4.1 UNDERSTANDING TODAY'S VIEW OF VALUE

In order to understand today's view of value, we must first understand how project management has changed over the years. Some critical changes are:

- Projects are highly complex and with greater acceptance of risks that may not be fully understood during project approval.
- There exists more uncertainty in the outcomes of the projects with no guarantee of value at the end of the project.
- We are pressed for speed-to-market irrespective of the risks.
- Enterprise project management methodologies now contain business processes as well as project management processes.

As these changes occurred, it became evident that value was now playing an important role in project selection and execution. Table 4-1 shows how our view of project-related value has changed.

TABLE 4-1 Changing View of Value	
PM 1.0	**PM 2.0**
All projects in the queue must be completed eventually	It does not matter if the project is completed if no business value is created
Being on time and within budget is the definition of success	Success is creating business value within the competing constraints
Being on time and within budget creates value	Time and cost are not the only characteristics of value
Good enterprise project management methodologies, when used correctly, will produce value	Methodologies are useful but cannot generate value by themselves
Customers want high-quality deliverables	Customers want deliverables that create business value; quality may be just one component of value
Value is measurable at the end of the project once the deliverables have been achieved	On some projects, value metrics can be established early on and tracked throughout the life of the project

These six statements lead us to believe that perhaps business value is now the dominating factor in the selection of a portfolio of projects. Project requestors must now clearly articulate the benefits and expected value in the project's business case or run the risk that the project will not be considered.

In today's world, customers make decisions to hire a contractor based upon the value they expect to receive and the price they must pay to receive this value. Actually it is more of a "perceived" value that may be based upon trade-offs on the attributes of the client's definition of value. *The client may perceive the value of your project to be used internally in their company or pass it on to their customers through their customer value management program.* If your organization does not or cannot offer recognized value to your clients and stakeholders, then you will not be able to extract value (i.e., loyalty) from them in return. Over time, they will defect to other contractors.

> The importance of value is clear. According to a study by the American Productivity and Quality Center (APQC): Although customer satisfaction is still measured and used in decision-making, the majority of partner organizations [used in this study] have shifted their focus from customer satisfaction to customer value.[1]

The definition of a project, as defined in Chapter 3, is "a set of values scheduled for sustainable realization." As a project manager, you must therefore establish metrics so that the client and stakeholders can track the value that you will be creating. Measuring and reporting customer value throughout the project are now a competitive necessity. If it is done correctly, it will build emotional bonds with your clients.

For years, the principles of value management have been applied to engineering and manufacturing activities, but only recently have the same principles been applied to project management. According to Venkataraman and Pinto[2]:

---

[1] "Customer Value Measurement: Gaining Strategic Advantage," The American Productivity and Quality Center (APQC), Houston, TX,1998, p. 8.

[2] Adapted from R. R. Venkataraman and J. K. Pinto, *Cost and Value Management in Projects*, Wiley, Hoboken, NJ, 2008; pp.164–165.

Value can be added to projects in several ways. These include providing greater levels of client satisfaction, maintaining acceptable levels of satisfaction while lowering resource expenditures, or some combination of the two. It is also possible to improve value by simultaneously increasing satisfaction and resources, provided that satisfaction increases more than the resources used to achieve it.

When managing projects for value, five fundamental concepts must be embraced:

- Concept 1: Projects derive their value from the benefits the organization accrues by achieving its stated goals.
- Concept 2: Project can be viewed as investments made by management.
- Concept 3: Project investors and sponsors tolerate risk.
- Concept 4: Project value is related to investment and risks.
- Concept 5: Value is a balance among the three key project elements: performance, resource usage, and risk.

Traditionally, business plans have tried to identify the benefits and resulting value expected from the project. The business plans were usually prepared by a business analyst (BA), and all of this was done prior to the project manager being assigned and brought on board the project. Unfortunately, once the project kicked off, the metrics being monitored and reported generally focused on time and cost rather than the value that the customer wanted or would be receiving. Value-based metrics were not reported because we simply did not know how to perform the value-based measurements.

Today, we can define a project as a collection of value scheduled for sustainable realization. With this definition, the role of the BA and the project manager are now coming together. As stated by Robert Wysocki[3]:

> Meeting time and cost constraints has very little to do with project success. Project success is measured in terms of business value expected compared to business value delivered. Both the PM and the BA should be making every effort to maximize business value for the time and cost invested. This puts the goals of the PM and the BA in alignment.

We can now define project success as the ability to achieve the desired value within the competing constraints imposed upon the project.

Today, with the growth of measurement management techniques, value-based metrics are a necessity for determining project success and are being considered as critical KPIs to be monitored and reported to the client. One contractor reports to their client on a monthly basis the amount of time left before the client will achieve the value that is expected, and the date may be beyond the end date of the project. However, many of the value-based metrics are still considered a measurement challenge.

4.2 VALUE MODELING

Surprisingly enough, numerous research on value has taken place over the past 20 years. With PM 1.0, the definition of value was:

$$\text{Value} = \frac{\text{quality}}{\text{cost}}$$

---

[3] R. K. Wysocki, *The Business Analyst Project Manager*, Wiley, Hoboken, NJ, 2011, p.2. The author provided an excellent discussion of the relationship between the project manager and the BA.

In other words, to provide additional value, one must either increase the quality or lower the cost. This equation implies that the only components of value are quality and cost. However, as value became important, researchers began to help define, measure, and report value using components other than just quality and cost. Some of the areas of research are:

- Value dynamics
- Value gap analysis
- Intellectual capital valuation
- Human capital valuation
- Economic value-based analysis
- Intangible value streams
- Customer value management/mapping
- Competitive value matrix
- Value chain analysis
- Valuation of IT projects
- Balanced scorecard

The output of this research has to be the creation of value models, many of which are directly applicable to project management and are used in PM 2.0. Some of the models are:

- Intellectual capital valuation
- Intellectual property scoring
- Balanced scorecard
- Future Value Management™
- Intellectual Capital Rating™
- Intangible value stream modeling
- Inclusive Value Measurement ™
- Value performance framework
- Value measurement methodology (VMM)

Value can appear in many forms as seen in Figure 4-1. For the purpose of this book, only economic values will be considered.

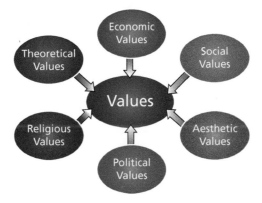

FIGURE 4-1   Types of values

4.3 VALUE AND LEADERSHIP CHANGES FOR PM 2.0

The importance of value has had a significant impact on the leadership style of project managers. Historically, with PM 1.0, project management leadership was perceived as the inevitable conflict between individual values and organizational values. Today, companies are looking for ways to get employees to align their personal values with the organization's values. Several books have been written on this subject and the best one, in this author's opinion, is *Balancing Individual and Organizational Values* by Ken Hultman and Bill Gellerman. Table 4-2 shows how our concept of value, as related to leadership, has changed over the years.[4]

As mentioned previously, each of us has our own interpretation of value. In a project environment, conflicts over the meaning of value can lead to poor decision making.

TABLE 4-2 Changing View of Value	
PM 1.0	**PM 2.0**
Mistrust of workers	Trusting workers
Lack of project management job descriptions	Job descriptions and competency models
Power and authority are important	Teamwork is important
Satisfying your superiors for performance reviews is essential	Stakeholder satisfaction is essential
Job security is critical	Project risk taking is necessary
Conform to the status quo	Innovation is a necessity
Project management methodologies focus on the rigidity of policies and procedures	Project managers must have flexibility in the use of the methodology
Internal competition exists	Competition is now between external companies
Reactive management	Proactive management
Formality is project execution	Informality in project execution
Adhere to the organizational bureaucracy	Bypassing bureaucracy may be necessary
Traditional project management education	Life-long project management education
Tactical thinking	Strategic thinking
Functional dedication	Project dedication
Meeting standards	Continuous improvements

---

[4] Adapted from K. Hultman and B. l. Gellerman, *Balancing Individual and Organizational Values*, Jossey-Bass/Pfeiffer, San Francisco, 2002, pp. 105–106.

Here is an example of how each group can interpret the value of a successfully completed project:

- Project manager
  - Accomplishment of objectives
  - Demonstration of creativity
  - Demonstration of innovation
- Team members
  - Achievement
  - Advancement
  - Ambition
  - Credentials
  - Recognition
- Organization
  - Continuous improvement
  - Learning
  - Quality
  - Strategic focus
  - Morality and ethics
  - Profitability
  - Recognition and image
  - Social consciousness
- Stakeholders
  - Organizational stakeholders: job security
  - Product/market stakeholders: quality performance and product usefulness
  - Capital markets: financial growth

There are several reasons why the role of the project manager and the accompanying leadership style have changed with PM 2.0. Some reasons are:

- We are now managing our business as though it is a series of projects.
- Project management is now viewed as a full-time profession.
- Project managers are now viewed as both business managers and project managers and are expected to make decisions in both areas.
- The value of a project is measured more so in business terms rather than in solely technical terms.
- Project management is now being applied to parts of the business that traditionally have not used project management.

This last bullet requires further comment. Project management works well for the "traditional" type of project, such as in PM 1.0, which includes the following:

- The time duration is 6–18 months.
- The assumptions are not expected to change over the duration of the project.
- Technology is known and will not change over the duration of the project.
- People that start on the project will remain through to completion.
- The statement of work (SOW) is reasonably well defined.

The newer types of projects that are more common in PM 2.0 are more nontraditional and have the following characteristics:

- The time duration is several years.
- The assumptions can and will change over the duration of the project.
- Technology will change over the duration of the project.
- People that approved the project may not be there at completion.
- The SOW is ill defined and subject to numerous changes.

The nontraditional types of projects have made it clear why traditional project management must change. There are three areas that necessitate changes:

- New projects have become:
  - Highly complex and with greater acceptance of risks that may not be fully understood during project approval
  - More uncertain in the outcomes of the projects and with no guarantee of value at the end
  - Pressed for speed-to-market irrespective of the risks
- The SOW is:
  - Not always well-defined, especially on long-term projects
  - Based upon possibly flawed, irrational, or unrealistic assumptions
  - Inconsiderate of unknown and rapidly changing economic and environmental conditions
  - Based upon a stationary rather than moving target for final value
- The management cost and control systems [enterprise project management (EPM) methodologies] focus on:
  - An ideal situation (as in the **PMBOK® Guide**)
  - Theories rather than the understanding of the workflow
  - Inflexible processes
  - Periodically reporting time at completion and cost at completion but not value (or benefits) at completion
  - Project continuation rather than canceling projects with limited or no value

Over the years, we have taken several small steps to plan for the use of project management on nontraditional projects, including the following:

- Project managers are provided with more business knowledge and are allowed to provide an input during the project selection process.
- Because of the above item, project managers are brought on board the project at the beginning of the initiation phase rather than the end of the initiation phase.
- Projects managers now seem to have more of an understanding of technology rather than a command of technology.

Defining success on a project in terms of business value has never been an easy task. The focus has always been the triple constraints or competing constraints. Today, with PM 2.0, we believe that there are four cornerstones for success defined in terms of value. This is shown in Figure 4-2.[5]

---

[5] A more detailed discussion of Figure 4-2 will appear in Chapter 8.

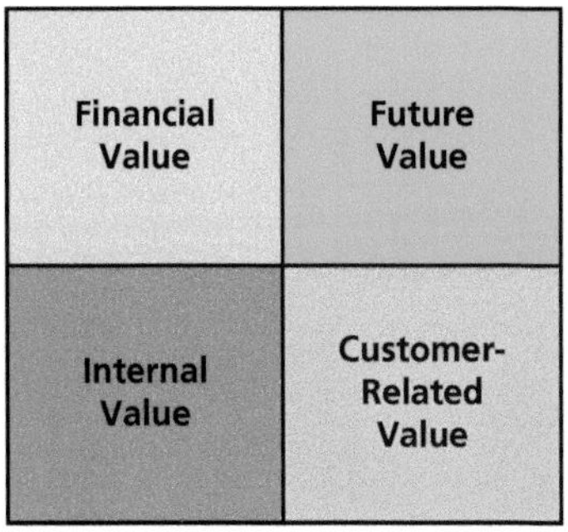

FIGURE 4-2 Categories of project value

The quadrants can be defined as follows:

Internal Value: The ability to have a continuous stream of successfully managed projects that create business value and by using an EPM methodology that undergoes continuous improvement on a regular basis. This could also include maintaining a scope change control process and building relationships within the firm.

Financial Value: The ability to create a long-term revenue stream that satisfied the financial needs of the stakeholders. Some practitioners also include here maintaining regulatory agency requirements [e.g., Occupational Safety and Health Administration (OSHA), U.S. Environmental Protection Agency (EPA)] and demonstrating ethical conduct.

Future Value: The ability to produce a stream of deliverables that will support the future business value needs of the firm. This includes maintaining technical superiority and process improvements.

Customer-Related Value: The ability to satisfy the business value needs of the customers over and over again to the point where you receive repeat business and the customers treat you as though you are a partner rather than a contractor or supplier. The focus is on customer satisfaction and building relationships.

Some projects can support more than one quadrant in Figure 4-2. For example, developing a product using your company's proprietary technology and having the product well-accepted in the marketplace can lead to financial, future and customer-related value. Another example would be the creation of an EPM methodology. Although the value of this can be regarded as internal value, the methodology can lead to a long-term relationship with customers and eventually financial value. Therefore, the success of the project and ultimate value achieved can impact more than one quadrant.

Figure 4-3 shows the relationship between the value achieved and the investment or discovery cost to achieve the values. As an example, in the short term, there may be a low discovery cost to improve the company's processes (i.e., internal values) but a potentially high cost to maintain long-term customer satisfaction. Long-term values support financial success and future success.

Achieving value can always be accomplished but perhaps at an unusually high cost. Executives must carefully weigh the cost of obtaining the desired value against what the value is actually worth. Spending $2 million to achieve a deliverable that provides

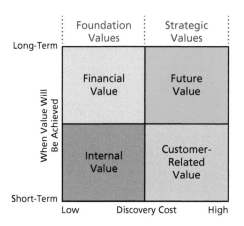

FIGURE 4-3   Modified categories of value

$1 million in value is not a good business decision. Of course, placing a dollar figure on value can be somewhat difficult, if not impossible to do, at least with reasonable accuracy.

In Figure 4-3 we also identified two major categories of values: foundation values and strategic values. The foundation values can be divided into internal values and financial values whereas the strategic values are more aligned with future values and customer-related values.

Foundation values are those values that must be achieved in the short term for the continuous operation of the firm on a day-to-day basis. This includes methodologies and processes to support ongoing activities. Cash flow is also needed to continue operations, so some activities that provide financial value are needed.

Strategic or innovation values are those values that must be achieved for the long-term survivability of the firm. This includes maintaining a strong list of clients, especially those that treat you as a potential partner, and having a pipeline of new projects that support the future products and services of the firm.

Deciding in which quadrant to place a given project is very subjective. Some projects can encompass more than one quadrant and possibly even all four quadrants. Because of this subjectivity, we can place certain projects or activities in the quadrant where maximum value will be or can be achieved. For example, consider a company that successfully improves their process for commercialization of a product. We could argue that it is an internal value because of process improvement; a financial value because it should allow us to increase our margins; a customer-related value because we can get the needed products to the customers quicker; or a future value because it is designed to improve potentially all products or services in the future. In our minds, this would fall under internal value.

4.4 VALUE-BASED TRADE-OFFS

Figures 4-4 and 4-5 illustrate how trade-offs usually occur with PM 1.0 and PM 2.0.

With traditional trade-offs as with PM 1.0, we tend to reduce performance to satisfy other requirements. The trade-off decisions are made by the project manager

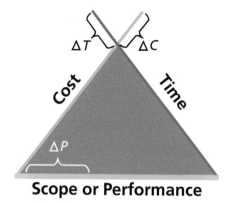

Scope or Performance

Note: △ = Deviations from the original plan

FIGURE 4-4 **PM 1.0 Trade-offs**

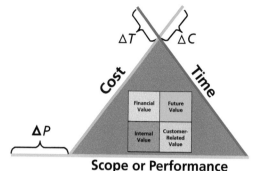

Scope or Performance

Note: △ = Deviations from the original plan

FIGURE 4-5 **PM 2.0 Trade-offs**

after getting approval by the project sponsor. The customer may or may not be involved in the trade-off decisions. With value-based trade-offs common to PM 2.0, we tend to increase performance in hopes of providing added value, and this tends to cause much larger cost overruns and schedule slippages than with traditional trade-offs. Projects managers generally do not have the sole authority for scope/performance increases or decreases with value trade-offs. These decisions may involve the entire governance committee. It is important to understand the need for the project manager to have a good business knowledge for PM 2.0 trade-offs.

For value-driven projects all or most of the stakeholders may need to be involved in trade-off decisions. This can create additional issues, such as:

- It may not be possible to get all of the stakeholders to agree on a value target during project initiation.
- Getting agreement on scope changes, extra costs, and schedule elongations is significantly more difficult the further along you are in the project.
- Stakeholders must be informed of this at project initiation and continuously briefed as the project progresses; that is, no surprises!

Trade-off conflicts among the stakeholders may occur. For example:

- During project initiation, conflicts among stakeholders are usually resolved in favor of the largest financial contributors.
- During execution, conflicts over future value are more complex, especially if major contributors threaten to pull out of the project.

For projects that have a large number of stakeholders, project sponsorship may not be effective with a single-person sponsor. As such, committee sponsorship may be necessary. Membership in the committee may include:

- Perhaps a representative from all stakeholder groups
- Influential executives
- Critical strategic partners and contractors
- Others based upon the type of value

Responsibilities for the sponsorship committee may include:

- Taking a lead role in the definition of the targeted value
- Taking a lead role in the acceptance of the actual value
- Ability to provide additional funding for added value opportunities
- Ability to assess changes in the enterprise environmental factors
- Ability to validate and revalidate the assumptions

Sponsorship committees may have significantly more expertise than the project manager in defining and evaluating the value in a project.

4.5 NEED FOR VALUE METRICS

With PM 2.0, it is now essential to create metrics that focus not only on business (internal) performance but also on performance toward customer satisfaction. If the customer cannot see the value in the project, then the project may be canceled and repeat business will not be forthcoming. Good value metrics can also result in less customer and stakeholder interference and meddling in the project. Value metrics can measure business case deterioration and whether value is being created or destroyed.

The need for an effective metrics management program (which will be discussed in Chapter 5) that focuses on value-based metrics is clear:

- There must be a customer/contractor/stakeholders agreement on how a set of metrics will be used to define success or failure; otherwise, you have just best guesses. Metrics that show that the value of the project is increasing as work progresses foster a better agreement and cooperation between all parties.
- Metric selection must cover the reality of the entire project; this can be accomplished with a set of core metrics supported by a value metric.
- A failure in effective metrics management, especially value metrics, can lead to stakeholder challenges and a loss of credibility.

We need to develop value-based metrics that can forecast stakeholder value, possibly shareholder value, and most certainly project value. Most models for creating this metric are highly subjective and are based upon assumptions that must be agreed upon upfront by all parties.

4.6 CREATING A VALUE METRIC

The ideal situation would be the creation of a single value metric that the stakeholders can use to make sure that the project is meeting or exceeding each stakeholder's expectation of value. The value metric can be a combination of traditional metrics and KPIs. Discussing the meaning of a single value metric may be more meaningful than discussing the individual components; the whole is often greater than the sum of the parts.

There must be support for the concept of creating a value metric. According to a global IT consulting company:

There has to be buy-in from both sides on the importance and substance of a value metric; it cannot be the latest fad—it has to be understood as a way of tracking the value of the project.

Typical criteria for a value metric may be:

- Every project will have at least one value metric or value KPI. In some industries, it may not be possible to use just one value metric.
- There may be a limit, such as five, for the number of value attributes that are part of the value metric. As we mature in the use of value metrics, the number of attributes can grow or be reduced. Not all attributes that we would like to have will be appropriate or practical.
- There will be weighting factors assigned to each component.
- The weighting factors and the component measurement techniques will be established by the project manager and the stakeholders at the onset of the project. There may be company policies on assigning the weighting factors.
- The target boundary boxes for the metrics will be established by the project manager and possibly the project management office (PMO). If a PMO does not exist, then there may be a project management committee taking responsibility for accomplishing this or it may be established by the funding organization.

To illustrate how this might work, assume that, for the IT projects you perform for your stakeholders, the attributes of the value metric will be:

- Quality (of the final software package)
- Cost (of development)
- Safety protocols (for security of information)
- Features (functionality)
- Schedule or timing (for delivery and implementation)

These attributes are agreed to by you, the client, and the stakeholders at the onset of the project. The attributes may come from your metric/KPI library or may be new attributes. Care must be taken to make sure that your organizational process assets can track, measure, and report on each attribute. Otherwise, additional costs may be incurred and these costs must be addressed up front so that they can be included in the contract price. Although not highly recommended, some of the attributes can change over the life of the project, especially on a complex project or one with significant unknowns and possibly a large number of scope changes.

Time and cost are generally attributes of every value metric. However, there may be special situations where neither time, nor cost, nor both are value metric attributes:

- The project must be completed by law, such as environmental projects, where failure to perform could result in stiff penalties.
- The project is in trouble but necessary, and we must salvage whatever value we can.
- We must introduce a new product to keep up with the competition regardless of the cost.
- Safety, aesthetic value, and quality are more important than time, cost, or scope.

Other attributes are almost always included in the value metric to support time and cost. However, time and/or cost may not be part of the value metric because the other attributes are considered more important than time and/or cost. Time and cost are still tracked and reported separately but are not part of the value metric report.

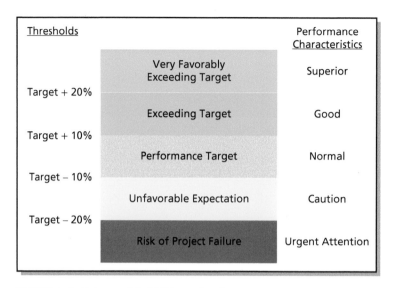

FIGURE 4-6 **Value metric/KPI boundary box**

The next step is to set up targets with thresholds for each attribute or component. This is shown in Figure 4-6. If the attribute is cost, then we might say that performing within ±10% of the cost baseline is normal performance. Performing at greater than 20% over budget could be disastrous, whereas performing at more than 20% below budget is superior performance. However, there are cases where a +20% variance could be good and a –20% variance could be bad.

The exact definition or range of the performance characteristics could be established by the PMO if company standardization is necessary or through an agreement with the client and the stakeholders. In any event, targets and thresholds must be established.

The next step is to assign value points for each of the cells in Figure 4-6, as shown in Figure 4-7. In this case, two value points were assigned to the cell labeled "performance target." The standard approach is to then assign points in a linear manner above and below the target cell. Nonlinear applications are also possible, especially when thresholds are exceeded.

In Table 4-3, weighting factors are assigned to each of the attributes of the value metric. As before, the weighting percentages could be established by the PMO or through an agreement with the client (i.e., funding organization) and the stakeholders. The use of the PMO might be for company standardization on the weighting factors. However, it sets a dangerous precedence when the weighting factors are allowed to change indiscriminately.

Now, we can multiply the weighting factors by the value points and sum them up to get the total value contribution. If all of the value measurements indicated that we were meeting our performance targets, then 2.0 would be the worth of the value metric. However, in this case, we are exceeding performance with regard to quality, safety, and schedule, and therefore the final worth of the value metric is 2.7. This implies that the stakeholders are receiving additional value that is most likely meeting or exceeding expectations.

	Performance Characteristics	Value Points
Very Favorably Exceeding Target	Superior	4
Exceeding Target	Good	3
Performance Target	Normal	2
Unfavorable Expectation	Caution	1
Risk of Project Failure	Urgent Attention	0

FIGURE 4-7 Value points for boundary box

TABLE 4-3 Value Metric Measurement

Value Component	Weighting Factor (%)	Value Measurement	Value Contribution
Quality	10	3	0.3
Cost	20	2	0.4
Safety	20	4	0.8
Features	30	2	0.6
Schedule	20	3	0.6
			Total = 2.7

There are still several issues that must be considered when using this technique:

- We must clearly define what is meant by normal performance. The users must understand what this means. Is this level actually our target level or is it the minimal acceptable level for the client? If it is our target level, then having a value below 2.0 might still be acceptable to the client if our target were greater than what the requirements asked for.
- The users must understand the real meaning of the value metric. When the metric goes from 2.0 to 2.1, how significant is that? Statistically, this is a 5% increase. Does it mean that that the value increased 5%? How can we explain to a layman the significance of such an increase and the impact on value?

Value metrics generally focus on the present and/or future value of the project and may not provide sufficient information as to other factors that may affect the health of the project. As an example, assume that the value metric is quantitatively assessed at a value of 2.7. From the customer's perspective, they are receiving more value than they anticipated. But other metrics may indicate that the project should be considered for termination. For example:

- The value metric is 2.7 but the remaining cost of development is so high that the product may be overpriced for the market.
- The value metric is 2.7 but the time to market will be too late.
- The value metric is 2.7 but a large portion of the remaining work packages have a very high critical risk designation.
- The value metric is 2.7 but significantly more critical assumptions are being introduced.
- The value metric is 2.7 but the project no longer satisfies the client's needs.
- The value metric is 2.7 but your competitors have introduced a product with a higher value and quality.

In Table 4-4, we reduced the number of features in the deliverable, which allowed us to improve quality and safety as well as accelerate the schedule. Since the worth of the value metric is 2.4, we are still providing additional value to the stakeholders.

In Table 4-5, we have added additional features as well as improving quality and safety. However, to do this, we have incurred a schedule slippage and a cost overrun.

TABLE 4-4 Value Metric with Reduction in Features

Value Component	Weighting Factor (%)	Value Measurement	Value Contribution
Quality	10	3	0.3
Cost	20	2	0.4
Safety	20	4	0.8
Features	30	1	0.3
Schedule	20	3	0.6
			Total = 2.4

TABLE 4-5 Value Metric with Improved Quality, Features, and Safety

Value Component	Weighting Factor (%)	Value Measurement	Value Contribution
Quality	10	3	0.3
Cost	20	1	0.2
Safety	20	4	0.8
Features	30	4	1.2
Schedule	20	1	0.2
			Total = 2.7

The worth of the value metric is now 2.7, which implies that the stakeholders are still receiving added value. The stakeholders may be willing to incur the added cost and schedule slippage because of the added value.

Whenever it appears that we may be over budget or behind schedule, we can change the weighting factors and overweigh those components that are in trouble. As an example, Table 4-6 shows how the weighting factors can be adjusted. Now, if the overall worth of the value metric exceeds 2.0 with the adjusted weighting factors, the stakeholders may still consider the continuation of the project. Sometimes, companies identify minimum and maximum weights for each component, as shown in Table 4-7. However, there is a risk that management may not be able to adjust to and accept weighting factors that can change from project to project or even during a project. Also, standardization and repeatability of the solution may disappear with changing weighting factors.

Companies are generally reluctant to allow project managers to change weighting factors once the project is underway and may establish policies to prevent unwanted changes from occurring. The fear is that the project manager may change the weighting factors just to make the project look good. However, there are situations where a change may be necessary:

- Customers and stakeholders are demanding a change in weighting factors possibly to justify the continuation of project funding.

TABLE 4-6 Changing the Weighting Factors

Value Component	Normal Weighting Factor (%)	Weighting Factors If We Have a Significant Schedule Slippage (%)	Weighting Factors If We Have a Significant Cost Overrun (%)
Quality	10	10	10
Cost	20	20	40
Safety	20	10	10
Features	30	20	20
Schedule	20	40	20

TABLE 4-7 Weighting Factor Ranges

Value Component	Minimal Weighting Value (%)	Maximum Weighting Value (%)	Nominal Weighting Value (%)
Quality	10	40	20
Cost	10	50	20
Safety	10	40	20
Features	20	40	30
Schedule	10	50	20

- The risks of the project have changed in downstream life-cycle phases and a change in weighting factors is necessary.
- As the project progresses, new value attributes are added into the value metric.
- As the project progresses, some value attributes no longer apply and must be removed from the value metric.
- The enterprise environment factors have changed requiring a change in the weighting factors.
- The assumptions have changed over time.
- The number of critical constraints has changed over time.

We must remember that project management metrics and KPIs can change over the life of a project and, therefore, the weighting factors for the value metric may likewise be susceptible to changes.

Sometimes, because of the subjectivity of this approach, when the information is presented to the client, we should include identification of which measurement technique was used for each target. This is shown in Table 4-8. The measurement techniques may be subject to negotiations at the beginning of the project.

The use of metrics and KPIs has been with us for decades, but the use of a value metric is relatively new. Therefore, failures in the use of this technique are still common and may include:

- Is not forward looking; the value metric focuses on the present rather than the future
- Does not go beyond financial metrics and, thus, fails to consider the value in knowledge gained, organizational capability, customer satisfaction, and political impacts
- Believing that value metrics (and the results) that other companies use will be the same for your company
- Not considering how the client and stakeholders define value
- Allowing the weighting factors to change too often to make the project's results look better

As with any new technique, additional issues always arise. Typical questions that we are now trying to answer in regard to the use of a value metric include:

- What if only three of the five components can be measured, for example, in the early life-cycle phases of a project?

TABLE 4-8 Weighting Factors and Measurement Techniques

Value Component	Weighting Factor (%)	Measurement Technique	Value Measurement	Value Contribution
Quality	10	Sampling techniques	3	0.3
Cost	20	Direct measurement	2	0.4
Safety	20	Simulation	4	0.8
Features	30	Observation	2	0.6
Schedule	20	Direct measurement	3	0.6

- In such a case where only some components can be measured, should the weighting factors be changed or normalized to 100% or left alone?
- Should the project be a certain percent complete before the value metric has any real meaning?
- Who will make decisions as to changes in the weighting factors as the project progresses through its life-cycle phases?
- Can the measurement technique for a given component change over each life-cycle phase or must it be the same throughout the project?
- Can we reduce the subjectivity of the process?

4.7 DISPLAYING VALUE METRICS IN A DASHBOARD

In Chapter 1, we stated that written reports will be replaced by dashboards. It is therefore imperative that the value metric be displayed correctly. This could very well be one of the most important characteristics of PM 2.0. Figure 4-8 illustrates how the value metric may appear on a dashboard. The value attributes and ratings in the table in the upper right corner reflect the values in the month of April. In January, the magnitude of the value metric was about 1.7. In April, the magnitude is 2.7.

The stakeholders can easily see the growth in value over the past four months. They can also see that four of the five attributes have increased their value over this time period, whereas the cost attribute appears to have diminished in value.

Eventually, as we become more knowledgeable in the use of value metrics, we may end up with a single value metric that can be obtained through an objective and automated process. In the near term, however, we can expect the value metric process to be more qualitative than quantitative and highly subjective based upon the value attributes that were selected.

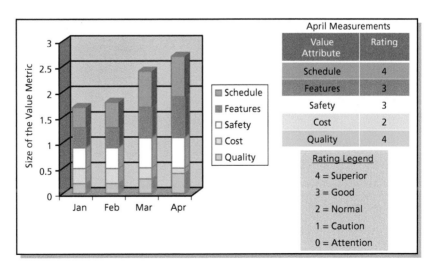

FIGURE 4-8   **Project value attributes**

4.8 SELECTING VALUE ATTRIBUTES

Selecting the right value attributes is not easy. The client may wish to select the value attributes based upon the best interests of either the client or the client's customers. The client may have specific value attributes that are of particular importance to the client and perhaps not directly meaningful to the project manager. A listing of these client preferred attributes appears in Table 4-9.

Even if the project manager and the client agree on the value attributes, there can be a different interpretation of the actual meaning of the attribute. As an example, Table 4-10 shows how various somewhat simple attributes can be interpreted differently.

TABLE 4-9 Client Preferred Attributes	
Value Metric Attribute	**Possible Competitive Advantage**
Deliverables produced	Efficiency
Product functionality	Innovation
Product functionality	Product differentiation
Support response time	Service differentiation
Staffing and employee pay grades	People differentiation
Quality	Quality differentiation
Action items in the system and for how long	Speed of problem resolution and decision making
Cycle time	Speed to market
Failure rates	Quality differentiation and innovation differentiation

TABLE 4-10 Interpretation of Value Attributes			
Generic Value Attributes	**Project Manager's Interpretation**	**Customer's Interpretation**	**Consumer's Interpretation**
Time	Project duration	Time to market	Delivery date
Cost	Project cost	Selling price	Purchase price
Performance	Quality	Functionality	Usability
Technology and scope	Meeting specifications	Strategic alignment	Safe buy and reliable
Satisfaction	Customer satisfaction	Consumer satisfaction	Esteem in ownership
Risks	No future business from this client	Loss of profits and market share	Need for technical support and risk of obsolescence

4.9 ADDITIONAL COMPLEXITIES WITH VALUE METRICS

Traditionally, business plans identified the benefits expected from the project. Today, With PM 2.0, portfolio management techniques require identification of the value as well as the benefits. However, conversion from benefits to value is not easy. With any new technique such as in PM 2.0, there are shortcomings in the conversion process from benefits to value that can make the conversion difficult. This is shown in Figure 4-9.

KPIs are metrics for assessing value. KPIs are quantifiable measures that are agreed to beforehand that reflect the success factors or the project or the firm. Value KPIs measure the progress to project and organizational goals. They measure strategic performance attributes and may be part of business intelligence (BI) systems. Value-reflective KPIs help us reduce uncertainty in order to make better decisions. Value-reflective KPIs lead to proactive project management.

With traditional project management, metrics are established by the EPM methodology and fixed for the duration of the project's life cycle. But with PM 2.0, metrics can change from project to project during each life-cycle phase and over time because of:

- The way the company defines value internally
- The way the customer and contractor jointly define success and value at project initiation
- The way the customer and contractor come to an agreement at project initiation as to what metrics should be used on a given project
- New or updated versions of tracking software
- Improvements to the EPM methodology and accompanying project management information system
- Changes in the enterprise environmental factors

Even with the best possible metrics, measuring value can be difficult. Some values are easy to measure while others are more difficult. The easy values to measure

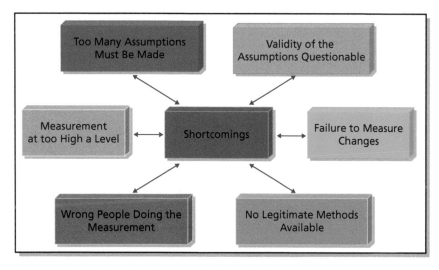

FIGURE 4-9 **Shortcomings in converting benefits to value**

are often called soft or tangible values, whereas the hard values are often considered intangible values.

The intangible elements are now considered by some to be more important than tangible elements. This appears to be happening on IT projects where executives are giving significantly more attention to intangible values where measurements are more of an art than a science. The critical issue with intangible values is not necessarily in the end result, but in the way that the intangibles were calculated. Tangible values are usually expressed quantitatively, whereas intangible values are expressed through a qualitative assessment.

The timing of value measurement is absolutely critical. During the life-cycle of a project, it may be necessary to switch back and forth from qualitative to quantitative assessment and, as stated previously, the actual metrics or KPIs can change as well. This type of flexibility is a necessity with PM 2.0. Certain critical questions must be addressed:

- When or how far along the project life-cycle can we establish concrete metrics, assuming it can be done at all?
- Can value be simply perceived and therefore no value metrics are required?
- Even if we have value metrics, are they concrete enough to reasonably predict actual value?
- Will we be forced to use value-driven project management practices on all projects or are there some projects where this approach is not necessary?
- Well-defined versus ill-defined requirements
- Strategic versus tactical projects
- Internal versus external clients

For some projects, assessing value at closure may be difficult. We must establish a time frame for how long we are willing to wait to measure the value or benefits from a project. This is particularly important if the actual value cannot be identified until sometime after the project has been completed. Therefore, it may not be possible to appraise the success of a project at closure if the true economic values cannot be realized until sometime in the future.

Some practitioners of value measurement question whether value measurement is better using boundary boxes instead of life-cycle phases. For PM 2.0 projects, the potential problems with life-cycle phases include:

- Metrics can change between phases and even during a phase.
- It may not be possible to account for changes in the enterprise environmental factors.
- Focus may be on the value at the end of the phase rather than the value at the end of the project.
- Team members may get frustrated not being able to quantitatively calculate value.

Boundary boxes, as shown in Figure 4-6, have some degree of similarity to statistical process control charts. Upper and lower strategic value targets are established. As long as the KPIs indicate that the project is still within the upper and lower value targets, the project's objectives and deliverables will not undergo any scope changes or trade-offs.

Most PM 2.0 projects will be more complex than PM 1.0 projects. As such, the projects must undergo value health checks to confirm that the project will make a contribution of value to the company. Value metrics, such as KPIs, indicate the current value. What is also needed is an extrapolation of the present into the future. Using tradition project management combined with the traditional EPM methodology, we can calculate the time at completion and the cost at completion. These are common terms that are helpful that appear in earned value measurement systems. But, as stated previously, being on time and within budget is no guarantee that the perceived value will be there at project completion.

Therefore, instead of using an EPM methodology which focuses on earned value measurement, we may need to modify our existing methodology to create a VMM which stresses the value variables. With VMM, time to complete and cost to complete are still used, but we introduce a new term, *value* (or benefits) *at completion*. Determination of value at completion must be done periodically throughout the project. However, periodic reevaluation of benefits and value at completion may be difficult because:

- There may be no reexamination process.
- Management is not committed and believes that the reexamination process is unreal.
- Management is overoptimistic and complacent with existing performance.
- Management is blinded by unusually high profits on other projects (misinterpretation).
- Management believes that the past is an indication of the future.

An assessment of value at completion can tell us if value trade-offs are necessary. Reasons for value trade-offs include:

- Changes in the enterprise environmental factors
- Changes in the assumptions
- Better approaches found, possibly with less risk
- Availability of highly skilled labor
- Breakthrough in technology

As stated previously, most value trade-offs are accompanied by an elongation of the schedule. Two critical factors that must be considered before schedule elongation takes place are:

- Elongating a project for the desired or added value may incur risks.
- Elongating a project consumes resources which may have already been committed to other projects in the portfolio.

Traditional tools and techniques may not work well on value-driven projects. The creation of a VMM may be necessary to achieve the desired results. A VMM can include the features of an EVMS and an EPM, as shown in Table 4-11. But additional variables must be included for the capturing, measurement, and reporting of value.

We must remember that PM 2.0 is just in the infancy stages. There are going to be significant challenges in order to reap the benefits. The process of measuring and reporting value will be one of the challenges.

Variable	EVMS	EPM	VMM
Time	√	√	√
Cost	√	√	√
Quality		√	√
Scope		√	√
Risks		√	√
Tangibles			√
Intangibles			√
Benefits			√
Value			√
Trade-offs			√

TABLE 4-11 Comparison of EVMS, EPM, and VMM

DISCUSSION QUESTIONS

The discussion questions are for classroom use to stimulate group thinking about PM 2.0. There are no right or wrong answers to most of the questions.

1. What is the definition of "value" with regard to a project?
2. Who defines the value of a project?
3. Can people working on a project, whether they are team members or governance personnel, have different definitions of value?
4. How do we determine the attributes of value?
5. Who determines the attributes of value?
6. How is project leadership impacted by our definition of value?
7. Is it possible that the actual value of a project cannot be determined until well after the project has been completed?
8. Is there a relationship between measurement techniques and value?
9. Can the value of a project be measured and displayed on a dashboard with a reasonable degree of accuracy?
10. In the future, is it possible that value metrics will replace the traditional metrics of time, cost, and scope?

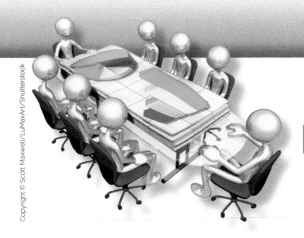

CHAPTER 5

GROWING IMPORTANCE OF METRICS WITH PM 2.0

5.0 INTRODUCTION

Project management metrics, which may be the most important component of PM 2.0, are part of a project management information system (PMIS) that contains all of the essential and supporting information for project approval, initiation, planning, scheduling, execution, monitoring and control, and closure. While EVMS is a critical component of the PMIS, today's PMIS contains significantly more metrics than just time and cost. The PMIS can provide significant benefits if designed properly, such as:

- Satisfying the information needs for the various stakeholders in a timely manner
- Providing the correct Information for informed decision making
- Having the correct amount of information, rather than too much or too little
- Lowering the cost of collecting the right information
- Providing information on how the project interacts with various initiatives that are part of the ongoing business
- Providing information on how one project interacts with other projects being supported by line managers
- Providing value to the company

A good PMIS, which is one of the characteristics of PM 2.0, can prevent projects from failing because of the derailment in project communications. PMIS also makes it easy for team members and functional managers to input the information necessary for effective status reporting more easily.

5.1 ENTERPRISE RESOURCE PLANNING

For several decades, PMISs were seen as report generators providing information on time, cost, and what work was left to do on the project. Time and cost were the two primary metrics that were tracked. Today, that has changed with PM 2.0. Metric management programs are the driving force behind PM 2.0.

Companies have come to the realization that everything they do in their company can be considered as a project. We are managing our business by projects. As such, decision makers have a need for information on both business and project management processes, and the two are now related. Also, we are now looking at significantly more metrics than just time and cost.

Information is the key to effective decision making. Companies have developed ERP systems which are enterprisewide information systems designed to coordinate all the resources, information, and tasks needed to complete various business and project processes. ERP supports supply chain management, finance and accounting, human resource management, and project management. PMIS is now part of ERP systems.

One of the most important reasons for integrating PMIS and ERP is capacity planning. Functional managers must supply resources to both projects and ongoing business activities. ERP systems are invaluable in this regard. Capacity planning is an essential activity in the portfolio selection of projects. As an example, the ERP system states that a given functional department has 15 employees available for work assignments and 10 of these workers are committed to ongoing work. The ERP system then relays information to the PMIS stating that the remaining 5 employees are available for project work assignments. The information can also contain the pay grade of the available workers in case-specific skill levels and must be available for the selection of certain projects.

5.2 NEED FOR BETTER PROJECT METRICS

With PM 2.0, project managers will have metrics at their fingertips.

Copyright © Scott Maxwell/Fotolia.

Metrics keep stakeholders informed as to the status of the project. Stakeholders must be confident that the correct metrics are being used and that the measurement portrays a clear and truthful representation of the status. At the beginning of a project, the project manager and the appropriate stakeholders must come to an agreement on which metrics to use and how measurements will be made. We are now using more metrics than just time and cost. This is partially due to the growth in PMIS and ERP technology as well as stakeholders now possessing a greater understanding about project management. Using Web 2.0 technology and combining it with PM 2.0 allow project team members and stakeholders to have immediate access to all metrics using mobile devices. The information will be displayed in real time.

Today, part of the project manager's new role in PM 2.0 is to understand what the critical metrics are that need to be identified and managed for the project to be viewed as a success by all of the stakeholders. Project managers have come to the realization that defining project-specific metrics and KPIs are joint ventures between the project manager, client, and stakeholders. Getting stakeholders to agree on the metrics is difficult, but it must be done as early as possible in the project.

Unlike financial metrics, project-based metrics can change during each life-cycle phase as well as from project to project. Therefore, the establishment and measurement of metrics may be an expensive necessity to validate the critical success factors (CSFs) and maintain customer satisfaction. Many people believe that the future will be metric-driven project management.

Although most companies use some type of metrics for measurement, they seem to have a poor understanding of what constitutes a metric, at least for use in project management. You cannot effectively manage a project without having metrics and accompanying measurement capable of providing you with complete or almost complete information. Therefore, the simplest definition of a metric is something that is measured. Consider the following facts about measurement and metrics:

- If it cannot be measured, then it cannot be managed.
- What gets measured gets done.
- You never really understand anything fully unless it can be measured.

If you cannot offer a stakeholder something that can be measured, then how can you promise that their expectations will be met? You cannot control what you cannot measure. Good metrics lead to proactive project management rather than reactive project management if the metrics are timely and informative.

PM 2.0 has shown us numerous benefits from the use of effective metrics management programs. Some benefits of using metrics are:

- Metrics tell us if we are hitting the targets/milestones, getting better, or getting worse.
- Metrics allow you to catch mistakes before they lead to other mistakes—early identification of issues.
- Good metrics lead to informed decision making, whereas poor or inaccurate metrics lead to bad management decisions.
- Good metrics can assess performance accurately.
- Metrics allow for proactive management in a timely manner.
- Metrics improve future estimating.
- Metrics improve performance for the future.
- Metrics make it easier to validate baselines and maintain the baselines with minimal disruptions.
- Metrics can more accurately assess success and failure.
- Metrics can improve client satisfaction.
- Metrics are a means of assessing the project's health.
- Metrics track the ability to meet the project's critical success factors.
- Good metrics allow the definition of project success to be made in terms of factors other than the traditional triple constraints.

The need for immediate access to the metric information should now be quite clear. The benefits of using metrics, even though there may be an initial cost of implementation, certainly outweighs the disadvantages. However, as will be seen later, there are hurdles that must be overcome to achieve the benefits.

While metrics are most frequently used to validate the health of a project, they can also be used to discover best practices in the project management processes. Capturing best practices and lessons learned is a necessity for long-term continuous improvement. Without effective use of metrics, companies could spend years trying to achieve sustained improvements. In this regard, metrics are a necessity not only to overcome some of the deficiencies of PM 1.0 but also to change some of the complexities of the project approval process, such as:

- Project approvals are often based upon insufficient information and poor estimating.
- Project approvals are based upon unrealistic ROI, net present value (NPV), internal rate of return (IRR), and payback period calculations.
- Project approvals are often based upon a best case scenario.
- The true time and cost requirements may be either hidden or not fully understood during the project approval process.

5.3 CAUSES FOR LACK OF SUPPORT FOR METRICS MANAGEMENT

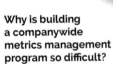

Even when a company recognizes the need for good metrics, metrics management programs can fail. The most common causes of failure are:

- Poor governance, especially by stakeholders
- Slow decision-making processes
- Overly optimistic project plans
- Trying to accomplish too much in too little time
- Poor project management practices and/or methodology
- Poor understanding of how the metrics will be used

Sometimes the failure of metrics management is due to poor stakeholder relations management. Typical issues that can lead to failure include the following:

- Failing to resolve disagreements among the stakeholders
- Failing to resolve mistrust among the stakeholders
- Failing to define CSFs
- Failing to get an agreement on the definition of project success
- Failing to get an agreement on the metrics needed to support the CSFs and the definition of success
- Failing to see if the CSFs are being met
- Failing to get an agreement on how to measure the metrics
- Failing to understand the metrics
- Failing to use the metrics correctly

Why is building a companywide metrics management program so difficult?

Copyright © Scott Maxwell/ LuMaxArt/Shutterstock

During the past few years, one of the drivers for effective metrics management has been the growth in complex projects. The larger and more complex the project, the greater the difficulty in measuring and determining success. Therefore, the larger and more complex the project, the greater the need for effective metrics.

But determining the metrics requires answering certain critical questions:

- Measurements
  - What should be measured?
  - When should it be measured?
  - How should it be measured?
  - Who will perform the measurement?

- Collecting information and reporting
  - Who will collect the information?
  - When will the information be collected?
  - When and how will the information be reported?

For many companies, answering these questions, especially on complex projects, was a challenge. As a result, metrics were often ignored because they were hard to define and collect.

Other reasons for the lack of support included:

- Metrics management was viewed as extra work and a waste of productive time.
- There was no guarantee that the correct metrics would be selected.
- If the wrong metrics were selected, then we were wasting time collecting the wrong data.
- Metrics management was considered as costly and the benefits did not often justify the cost.
- Metrics were considered expensive and useless.

Metric management is often seen as an add-on to the existing work of the project team. But without these metrics, we often focus on reactive rather than proactive management. The result is a focus on the completion of individual work packages rather than a focus on completion of the business solution for the client.

Everyone understands the value in using metrics. But there is still the inherent fear among team members that metrics will be seen as "Big Brother Is Watching You!" Employees will not support a metrics management effort that looks like a spying machine.

5.4 CHARACTERISTICS OF A METRIC

What are the characteristics of a good metric?

Copyright © Scott Maxwell/Fotolia

Establishing metrics requires knowledge of the basic requirements for a metric to exist. Metrics require:

- A need or purpose
- A target, baseline, or reference point
- A means of measurement
- A means of interpretation
- A performance-reporting structure

There are certain basic characteristics that a metric should possess. These include:

- Has a need or a purpose
- Provides useful information
- Focuses toward a target
- Can be measured with reasonable accuracy
- Reflects the true status of the project
- Supports proactive management
- Assists in assessing the likelihood of success or failure
- Accepted by the stakeholders as a tool for informed decision making

Unlike business environments which are long term, project environments are much shorter and therefore more susceptible to changing metrics. In a project environment, metrics can change from project to project, during each life-cycle phase, and at any time because of:

- The way the company defines value internally
- The way the customer and the contractor jointly define success and value at project initiation
- The way the customer and contractor come to an agreement at project initiation as to what metrics should be used on a given project
- New or updated versions of tracking software
- Improvements to the EPM methodology and accompanying PMIS
- Changes in the enterprise environmental factors
- Changes in the project's business case assumptions

Metrics can be classified. As an example, below are seven types of metrics or metric indicators that could appear in a metrics library:

- Quantitative metrics (planning dollars or hours as a percentage of total labor)
- Practical metrics (improved efficiencies)
- Directional metrics (risk ratings getting better or worse)
- Actionable metrics (affect change as the number of unstaffed hours)
- Financial metrics (e.g., profit margins, ROI)
- Milestone metrics (number of work packages on time)
- End result or success metrics (customer satisfaction and follow-on work)

5.5 METRICS SELECTION

Finding a compromise on the correct number of metrics is not easy, but we must determine how many metrics are actually needed for a particular project.

- With too many metrics:
  - Metric management steals time from other work.
  - We end up providing too much information to stakeholders such that they cannot determine what information is critical.
  - We end up providing information that has limited value.
- With too few metrics:
  - Not enough critical information is provided.
  - Informed decision making becomes difficult.

Why is it so difficult to select the right metrics for a project?

There are certain ground rules we can establish as part of the metric selection process:

- Make sure that the metrics are worth collecting.
- Make sure that we use what we collect.
- Make sure that the metrics are informative.
- Train the team in the use and value of metrics.

Selecting metrics is a lot easier when you have competent baselines from which to make measurements. It is very difficult or even impossible to use metrics management effectively when the baselines undergo continuous transformation. For work that has not been planned yet, benchmarks and standards can be used instead of baselines.

Metrics by themselves are just numbers or trends resulting from measurements. Metrics have no real value unless they can be properly interpreted by the stakeholders or subject matter experts and a corrective plan, if necessary, can be developed. It is important to know who will benefit from each metric. The level of importance can vary from stakeholder to stakeholder.

There are several questions that can be addressed during metric selection:

- How knowledgeable are the stakeholders in project management?
- How knowledgeable are the stakeholders in metrics management?
- Do we have the necessary organizational process assets for metric measurements?
- Will the baselines and standards undergo transformations during the project?

There are two additional factors that must be considered when selecting metrics. First, there is a cost involved in performing the measurements, and based upon the frequency of the measurements, the costs can be quite large. Second, we must recognize that metrics need to be updated. Metrics are like best practices; they age and may no longer provide the value or information that was expected. There are several reasons therefore for periodically reviewing the metrics:

- Customers may desire real-time reporting rather than periodic reporting, thus making some metrics inappropriate.
- The cost and complexity of the measurement may make a metric inappropriate for use.
- The metric does not fit well with the organizational process assets available for an accurate measurement.
- Project funding limits may restrict the number of metrics that can be used.

In reviewing the metrics, there are three possible outcomes:

- Update the metric.
- Leave the metric as is but possibly put it on hold until the next review.
- Retire the metric from use.

Finally, metrics should be determined after the project is selected and approval is obtained. Selecting a project based upon available or easy-to-use metrics often results in either the selection of the wrong project or metrics that provide useless data.

5.6 KEY PERFORMANCE INDICATORS

As stated in the previous section, part of the project manager's role is to understand what the critical metrics are that need to be identified, measured, reported, and managed such that the project will be viewed as a success by all of the stakeholders, if possible. The term "metric" is generic whereas a "KPI" is specific. KPIs serve as early warning signs that, if an unfavorable condition exists and is not addressed, the results could be poor. KPIs and metrics can be displayed in dashboards, scorecards, and reports. KPIs will serve as a critical component of PM 2.0.

Defining the correct metrics or KPIs are joint ventures between the project manager, client, and stakeholders and are a necessity in order to get stakeholder agreement. One of the keys to a successful project is the effective and timely management of information. This includes the KPIs. KPIs give us information for making informed decisions by reducing uncertainty.

Getting stakeholder agreement on the KPIs is difficult, but it is a necessity. All of the stakeholders must be on the same page. If you provide the stakeholders with 50 metrics to select from, they will somehow justify the need for all 50 of them. If you show them 100 metrics, they will find a reason why all 100 should be reported. The hard part is to select from the metrics library those critical metrics which can function as KPIs.

After selecting the KPIs, there are three additional questions that stakeholders need answered:

- Is there a system in place for measuring the KPIs?
- Is there a system in place for communicating the results?
- Are there processes in place for helping us take corrective action when needed?

For years, metrics and KPIs were used primarily as part of business intelligence techniques. When applied to projects, KPIs answer the question, "What is really important for different stakeholders to monitor on the project?" In business, once a KPI is established, it becomes difficult to change as enterprise environmental factors change for fear that historical comparison data will be lost. But benchmarking industry KPIs is still possible because the KPIs are long term. In project management, because of the uniqueness of projects, benchmarking is more complex because of the relatively short life span of the KPIs.

Need for KPIs

Most often, the items that appear in the dashboards and reports are elements that both customers and project managers track. These items are referred to as KPIs. According to Eckerson[1]:

> A KPI is a metric measuring how well the organization or an individual performs an operational, tactical or strategic activity that is critical for the current and future success of the organization.

Although Eckerson's comment is more appropriate for business-oriented rather than project-oriented metrics, the application to a project environment still exists. KPIs are high-level snapshots of how a project is progressing toward predefined targets. Some people confuse a KPI with leading indicators. A leading indicator is actually a KPI that measures how the work you are doing now will affect the future. KPIs can be treated as indicators but not necessarily leading indicators.

While some metrics may appear as leading indicators, care must be taken as to how they are interpreted. The misinterpretation of a metric or the mistaken belief that a metric is a leading indicator can lead to faulty conclusions.

KPIs are critical components of all EVMSs. Terms such as cost variance, schedule variance, schedule performance index, cost performance index, and time/cost at completion

[1] W. W. Eckerson, *Performance Dashboards: Measuring, Monitoring and Managing Your Business*, Wiley, Hoboken, NJ, 2006, p . 294.

are actually KPIs if used correctly but not always referred to as such. The need for these KPIs is simple: What gets measured gets done! If the goal of a performance measurement system is to improve efficiency and effectiveness, then the KPI must reflect controllable factors. There is no point in measuring an activity if the users cannot change the outcome.

For more than four decades, the only KPIs we looked at were time and cost or derivatives of time and cost. Today, we realize that true project status cannot be measured from just time and cost alone. As such, the need for additional KPIs has grown. Typical KPIs that project managers may use include:

- Percent of work packages adhering to the schedule
- Percent of work packages adhering to the budget
- Number of assigned resources versus planned resources
- Percent of actual versus planned baselines completed to date
- Percent of actual versus planned best practices used
- Project complexity factor
- Time to achieve value
- Customer satisfaction ratings
- Number of critical assumptions made
- Percent of critical assumptions that have changed
- Number of cost revisions
- Number of schedule revisions
- Number of scope change review meetings
- Number of critical constraints
- Percent of work packages with a critical risk designation
- Net operating margins
- Grade levels of assigned resources versus planned resources

Project managers must explain to the stakeholders the differences between metrics and KPIs and why only the KPIs should be reported on dashboards. As an example, metrics focus on the completion of work packages, achievement of milestones, and accomplishment of performance objectives. KPIs focus on future outcomes and this is the information stakeholder's need for informed decision making. Neither metrics nor KPIs can truly predict that the project will be successful, but KPIs provide more accurate information on what might happen in the future if the existing trends continue. Both metrics and KPIs provide useful information, but neither can tell you what action to take or whether a distressed project can be recovered. A metric can serve as a traditional metric on one project but as a KPI on another project. The decision on what is or is not a KPI is very subjective and often based upon the type of project and the information to be provided.

Once the stakeholders understand the need for correct KPIs, other questions must be discussed, including:

- How many KPIs are needed?
- How often should they be measured?
- What should be measured?
- How complex will the KPI become?
- Who will be accountable for the KPI (i.e., the KPI owner)?
- Will the KPI serve as a benchmark?

We stated previously that what gets measured gets done, and it is through measurement that a true understanding of the information is obtained. If the goal of a metric measurement system is to improve efficiency and effectiveness, then the KPI must reflect controllable factors. There is no point in measuring an activity or a KPI if the users cannot change the outcome. Such KPIs would not be acceptable to stakeholders.

Using KPIs

Although most companies use metrics and perform measurement, they seem to have a poor understanding of what constitutes a KPI for projects and how they should be used. Some general principles are:

- KPIs are agreed to beforehand and reflect the CSFs on the project.
- KPIs indicate how much progress has been made toward the achievement of the project's targets, goals, and objectives.
- KPIs are not performance targets.
- The ultimate purpose of a KPI is the measurement of items directly relevant to performance and to provide information on controllable factors appropriate for decision making such that it will lead to positive outcomes.
- Good KPIs drive change but do not prescribe a course of action. They indicate how close you are to a target but do not tell you what must be done to correct deviations from the target.
- KPIs assist in the establishment of objectives to be targeted with the ultimate purpose of either adding value to the project or achieving the prescribed value.

Some people argue that the high-level purposes of a KPI are to encourage effective measurement. In this regard, the three high-level purposes are:

- Measurements that lead to motivation of the team
- Measurements that lead to compliance with use of organizational process assets and alignment to business objectives
- Measurements that lead to performance improvements and the capturing of lessons learned and best practices

Some companies post KPI information on bulletin boards, in the company cafeteria, on the walls of conference rooms, or in company newsletters as a means of motivating the organization by showing progress toward that target. However, unfavorable KPIs can have an adverse effect on morale.

Anatomy of a KPI

Some metrics, such as project profitability, can tell us if things look good or bad but do not necessarily provide meaningful information on what we must do to improve performance. Therefore, a typical KPI must do more than just function as a metric. If we dissect the KPIs, we will see the following:

KEY = a major contributor to the success or failure of the project. A KPI metric is therefore only "key" when it can make or break the project.

PERFORMANCE = a metric that can be measured, quantified, adjusted, and controlled. The metric must be controllable to improve performance.

INDICATOR = reasonable representation of present and future performance.

A KPI is part of a measurable objective. Defining and selecting the KPIs are much easier if you define the CSFs first. KPIs should not be confused with CSFs. CSFs are things that must be in place to achieve an objective. A KPI is not a CSF but may provide a leading indication that the CSF can be met.

Selecting the right KPIs and the right number of KPIs will:

- Allow for better decision making
- Improve performance on the project
- Help identify problem areas faster
- Improve customer–contractor–stakeholder relations

David Parmenter[2] defines three categories of metrics:

- **Results Indicators (RIs)**: What have we accomplished?
- **Performance Indicators (PIs)**: What must we do to increase or meet performance?
- **Key Performance Indicators (KPIs)**: What are the critical performance indicators that can drastically increase performance or accomplishment of the objectives?

Most companies use an inappropriate mix of these three and label them as KPIs. Having too many KPIs can slow down projects due to excessive measurements and reporting requirements. Too many can also blur one's vision on actual performance. Too few can likewise cause delays because of the lack of critical information. Typically, we end up with too many rather than too few KPIs.

The number of KPIs can vary from project to project and may be impacted by the number of stakeholders. Some people select the number of KPIs based upon the Pareto principle, which states that 20% of the total indicators will impact 80% of the project. David Parmenter states that the 10/80/10 rule is usually applied when selecting the number of KPIs[3]:

- RIs: 10
- PIs: 80
- KPIs: 10

Typically, between 6 and 10 KPIs are standard. Factors influencing the number of KPIs include:

- The number of information systems that the project manager uses (i.e. 1, 2, or 3)
- The number of stakeholders and their reporting requirements
- The ability to measure the information
- The organizational process assets available to collect the information
- The cost of measurement and collection
- Dashboard reporting limitations

---

[2] D. Parmenter, *Key Performance Indicators*, Wiley, Hoboken, NJ, 2007, p. 1.
[3] Ibid., p. 9

KPI Characteristics

The literature abounds with articles defining the characteristics of metrics and KPIs. All too often, authors use the "SMART" rule as a means of identifying the characteristics:

S = Specific: clear and focused toward performance targets or a business purpose

M = Measurable: can be expressed quantitatively

A = Attainable: the targets are reasonable and achievable

R = Realistic or relevant: the KPI is directly pertinent to the work done on the project

T = Time based: the KPI is measurable within a given time period

The SMART rule was originally developed for establishing meaningful objectives for projects and later adapted to the identification of metrics and KPIs. While the use of the SMART rule does have some merit, its applicability to KPIs is questionable.

The most important attribute of a KPI may be that it is actionable. If the trend of the metric is unfavorable, then the users should know what action is necessary to correct the unfavorable trend. The user must be able to control the outcome. This is a weakness when using the SMART rule to select KPIs.

Wayne Eckerson has developed a more sophisticated set of characteristics for KPIs. The list is more appropriate for business-oriented KPIs than project-oriented KPIs but can be adapted for project management usage. Table 5-1 shows Eckerson's 12 characteristics.[4]

Business or financial metrics are usually the results of many factors and it therefore may be difficult to isolate what must be done to implement change. For project-oriented KPIs, the following six characteristics may very well be sufficient[5]:

Predictive: Able to predict the future of this trend.

Measurable: Can be expressed quantitatively.

Actionable: Triggers changes that may be necessary for corrective action.

Relevant: The KPI is directly related to the success or failure of the project.

Automated: Reporting minimizes the chance of human error.

Few in Number: Only what is necessary.

Sometimes KPIs are categorized according to what they are intended to indicate, similar to the metrics categories discussed in the previous section:

- Quantitative KPIs: numerical values
- Practical KPIs: interfacing with company processes
- Directional KPIs: getting better or worse
- Actionable KPIs: lead to a change
- Financial KPIs: performance measurements

Another means of classification might be leading or lagging indicators or KPIs:

- Lagging KPIs measure past performance.
- Leading KPIs measure drivers for future performance.

Most dashboards have a compromise of both leading and lagging metrics.

---

[4] W. W. Eckerson, *Performance Dashboards: Measuring, Monitoring and Managing Your Business*, Wiley, Hoboken, NJ, 2006, p. 201.

[5] For additional information, see H. Kerzner, *Project Management Metrics, KPIs and Dashboards*, Wiley and International Institute for Learning, New York, 2013, Chapter 4.

TABLE 5-1 Twelve Characteristics of Effective KPIs
1. **Aligned**. KPIs are always aligned with corporate strategy and objectives.
2. **Owned.** Every KPI is "owned" by an individual or group on the business side who is accountable for its outcome.
3. **Predictive.** KPIs measure drivers of business value. Thus, they are "leading" indicators of performance desired by the organization.
4. **Actionable**. KPIs are populated with timely, actionable data so users can intervene to improve performance before it is too late.
5. **Few in number.** KPIs should focus users on a few high-value tasks, not scatter their attention and energy on too many things.
6. **Easy to understand.** KPIs should be straightforward and easy to understand, not based on complex indexes that users do not know how to influence directly.
7. **Balanced and linked.** KPIs should balance and reinforce each other, not undermine each other and suboptimize processes.
8. **Trigger changes.** The act of measuring a KPI should trigger a chain reaction of positive changes in the organization, especially when it is monitored by the CEO.
9. **Standardized.** KPIs are based on standard definitions, rules, and calculations, so they can be integrated across dashboards throughout the organization.
10. **Context driven.** KPIs put performance in context by applying targets and thresholds to performance so users can gauge their progress over time.
11. **Reinforced with incentives.** Organizations can magnify the impact of KPIs by attaching compensation or incentives to them. However, they should do this cautiously, applying incentives only to well-understood and stable KPIs.
12. **Relevant.** KPIs gradually lose their impact over time, so they must be periodically reviewed and refreshed.

KPI Failures

There are several reasons why the use of KPIs often fails on projects. Some of the reasons are:

- People believe that the tracking of a KPI ends at the first-line manager level.
- The actions needed to regulate unfavorable indications are beyond the control of the employees doing the monitoring or tracking.
- The KPIs are not related to the actions or work of the employees doing the monitoring.
- The rate of change of the KPIs is too slow, thus making them unsuitable for managing the daily work of the employees.
- Actions needed to correct unfavorable KPIs take too long.
- Measurement of the KPIs does not provide enough meaning or data to make them useful.
- The company identifies too many KPIs to the point where confusion reigns among the people doing the measurements.

Years ago, the only metrics that some companies used were those identified as part of the EVMS. The metrics generally focused only on time and cost and neglected metrics related to business success as opposed to project success. As such, the measurement metrics were the same on each project and the same for each life-cycle phase. Today, metrics can change from phase to phase and from project to project. The hard part is obviously deciding upon which metrics to use. Care must be taken that whatever metrics are established does not end up comparing apples and oranges. Fortunately, there are several good books in the marketplace that can assist in identifying proper or meaningful metrics.[6]

5.7 DASHBOARDS AND SCORECARDS

In our attempt to go to paperless project management, which is an important characteristic of PM 2.0, emphasis is being given to visual displays such as dashboard and scorecards. Executives and customers desire a visual display of the most critical project performance information in the least amount of space. Simple dashboard techniques, such as traffic light reporting, can convey critical performance information. For example:

- **Red Traffic Light**: A problem exists which may affect time, cost, quality, or scope. Sponsorship involvement is necessary.
- **Yellow or Amber Light**: This is a caution. A potential problem may exist, perhaps in the future if not monitored. The sponsor is informed but no action by the sponsor is necessary at this time.
- **Green Light**: Work is progressing as planned. No involvement by the sponsor is necessary.

While a traffic light dashboard with just three colors is most common, some companies use many more colors. The IT group of a retailer had an eight-color dashboard for IT projects. An amber color meant that the targeted end date had passed and the project was still not complete. A purple color meant that this work package was undergoing a scope change that could have an impact on the triple constraint.

Some people confuse dashboards with scorecards. There is a difference between dashboards and scorecards. According to Eckerson[7]:

- Dashboards are visual display mechanisms used in an *operationally* oriented performance measurement system that measure performance against targets and thresholds using right-time data.
- Scorecards are visual displays used in a *strategically* oriented performance measurement system that chart progress toward achieving strategic goals and objectives by comparing performance against targets and thresholds.

Both dashboards and scorecards are visual display mechanisms within a performance measurement system that convey critical information. The primary difference

---

[6] Three books that provide examples of metric identification are: P. F. Rad and G. Levin, *Metrics for Project Management*, Management Concepts, Vienna, VA, 2006; M. Schnapper and S. Rollins, *Value-Based Metric for Improving Results*, J. Ross Publishing, Ft. Lauderdale, FL, 2006; D. W. Hubbard, *How to Measure Anything*, 3rd. ed., Wiley, Hoboken, NJ, 2014.

[7] W. W. Eckerson, *Performance Dashboards: Measuring, Monitoring and Managing Your Business*, Wiley, Hoboken, NJ, 2006, pp. 293, 295. Chapter 12 provides an excellent approach to designing dashboard screens.

Feature	Dashboard	Scorecard
Purpose	Measures performance	Charts progress
Users	Supervisors, specialists	Executives, managers, and staff
Updates	Right-time feeds	Periodic snapshots
Data	Events	Summaries
Display	Visual graphs, raw data	Visual graphs, comments

TABLE 5-2 Comparing Features

between dashboards and scorecards is that dashboards monitor operational processes such as those used in project management, whereas scorecards chart the progress of tactical goals. Table 5-2 and the description following it show how Eckerson compares the features of dashboards and scorecards.[8]

Dashboards. Dashboards are more like automobile dashboards. They let operational specialists and their supervisors monitor events generated by key business processes. But unlike automobiles, most business dashboards do not display events in "real time" as they occur; they display them in "right time" as users need to view them. This could be every second, minute, hour, day, week, or month depending on the business process, its volatility, and how critical it is to the business. However, most elements on a dashboard are updated on an intraday basis, with latency measured in either in minutes or hours.

Dashboards often display performance visually, using charts or simple graphs, such as gauges and meters. However, dashboard graphs are often updated in place, causing the graph to "flicker" or change dynamically. Ironically, people who monitor operational processes often find the visual glitz distracting and prefer to view the data in its original form, as numbers or text, perhaps accompanied by visual graphs.

Scorecards. Scorecards, on the other hand, look more like performance charts used to track progress toward achieving goals. Scorecards usually display monthly snapshots of summarized data for business executives who track strategic and long-term objectives or daily and weekly snapshots of data for managers who need to chart the progress of their group or project toward achieving goals. In both cases, the data are fairly summarized so users can view their performance status at a glance.

Like dashboards, scorecards also make use of charts and visual graphs to indicate performance state, trends, and variance against goals. The higher up the users are in the organization, the more they prefer to see performance encoded visually. However, most scorecards also contain (or should contain) a great deal of textual commentary that interprets performance results, describes action taken, and forecasts future results.

Summary. In the end, it does not really matter whether you use the term dashboard or scorecard as long as the tool helps to focus users and organizations on what really matters. Both dashboards and scorecards need to display critical performance information on a single screen so users can monitor results at a glance.

[8] Ibid., p. 13.

TABLE 5-3 **Three Types of Performance Dashboards**			
	Operational	**Tactical**	**Strategic**
Purpose	Monitor operations	Measure progress	Execute strategy
Users	Supervisors, specialists	Managers, analysts	Executives, managers, staff
Scope	Operational	Departmental	Enterprise
Information	Detailed	Detailed/summary	Detailed/summary
Updates	Intraday	Daily/weekly	Monthly/quarterly
Emphasis	Monitoring	Analysis	Management

Although the terms are used interchangeably, most project managers prefer to use dashboards and/or dashboard reporting. Eckerson defines three types of dashboards as shown in Table 5-3 and the description that follows[9]:

Operational dashboards monitor core operational processes and are used primarily by front-line workers and their supervisors who deal directly with customers or manage the creation or delivery of organizational products and services. Operational dashboards primarily deliver detailed information that is only lightly summarized. For example, an online Web merchant may track transactions at the product level rather than the customer level. In addition, most metrics in an operational dashboard are updated on an intraday basis, ranging from minutes to hours depending on the application. As a result, operational dashboards emphasize monitoring more than analysis and management.

Tactical dashboards track departmental processes and projects that are of interest to a segment of the organization or a limited group of people. Managers and business analysts use tactical dashboards to compare performance of their areas or projects, to budget plans, forecasts, or last period's results. For example, a project to reduce the number of errors in a customer database might use a tactical dashboard to display, monitor, and analyze progress during the previous 12 months toward achieving 99.9% defect-free customer data by 2007.

Strategic dashboards monitor the execution of strategic objectives and are frequently implemented using a balanced scorecard approach, although total quality management, Six Sigma, and other methodologies are used as well. The goal of a strategic dashboard is to align the organization around strategic objectives and get every group marching in the same direction. To do this, organizations roll out customized scorecards to every group in the organization and sometimes to every individual as well. These "cascading" scorecards, which are usually updated weekly or monthly, give executives a powerful tool to communicate strategy, gain visibility into operations, and identify the key drivers of performance and business value. Strategic dashboards emphasize management more than monitoring and analysis.

There are three critical steps that must be considered when using dashboards; (1) the target audience for the dashboard, (2) the type of dashboard to be used, and

---

[9] Ibid., pp. 17–18.

(3) the frequency with which the data will be updated. Some project dashboards focus on the KPIs that are part of earned value measurement. These dashboards may need to be updated daily or weekly. Dashboards related to the financial health of the company may be updated weekly or quarterly.

5.8 BUSINESS INTELLIGENCE

Corporations have been using the concept of business intelligence (BI) for more than two decades. In recent years, business intelligence applications have been replaced by strategic intelligence (SI) applications supported by more meaningful metrics and many of the principles of PM 2.0. Both applications are designed around the monitoring and surveillance of business metrics. According to Corine Cohen[10]:

> The general surveillance field covers notions of watch, scanning, intelligence, competitive intelligence, vigilance, business intelligence, economic intelligence, economic and strategic intelligence, etc...
>
> SI is defined here as a formalized process of research, collection, information processing and distribution of knowledge useful to strategic management. Besides its information function, the main goals of SI are to anticipate environmental threats and opportunities (anticipatory function), help in strategic decision making and improve competitiveness and performance of the organization. It requires an organizational network structure, and human technical and financial resources.
>
> A distinction must therefore be made between Strategic Watch and SI. SI goes beyond Strategic Watch with its proactivity and its deeper involvement in the strategic decision process. Watch can (must) indicate the impacts of a detected event for example. However, it becomes intelligence when it produces recommendations and provides instructions to the recipient (all the more so when it implements them).

BI and SI applications have taught us that the way we try to monitor and control projects must change. In a project management environment, BI would be represented by metrics and SI would be represented by KPIs. Key performance indicators are the "strategic" metrics that provide us with the critical information for informed decision making. BI metrics are simply monitoring metrics whereas SI metrics, or KPIs, provide information on the future rather than just the present and indicate changes that may be necessary. Since project managers today and in the future will become more business-oriented managers, the relationship between metrics and BI and SI will become more important.

5.9 GROWTH IN DASHBOARD INFORMATION SYSTEMS

Perhaps every project manager's greatest fear with PM 2.0 is that stakeholders may request that all metrics that are in the metrics library be displayed on dashboard screens. With 50–100 metrics in the library, this could create information-sharing headaches.

[10] C. Cohen, *Business Intelligence*, Wiley, Hoboken, NJ, and ISTE Publishers, Washington, DC, 2009, p. xiii.

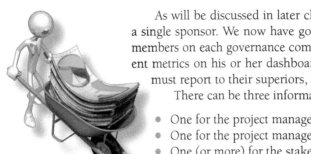

As will be discussed in later chapters, project governance is no longer handled by a single sponsor. We now have governance committees and there may be 10 or more members on each governance committee. Each committee member may request different metrics on his or her dashboards based upon their interests, the information they must report to their superiors, and the decisions they must make.

There can be three information systems on a project:

- One for the project manager
- One for the project manager's parent company
- One (or more) for the stakeholders, the governance committee, and the client

How many dash-boards do we need?

Copyright © Scott Maxwell/Fotolia

There can be a different set of metrics and KPIs in each of these information systems. It is important to avoid "metric mania," which is based upon the misguided belief that all metrics must be displayed. If at all possible, the project manager should try to limit the number of dashboards created.

5.10 SELECTING AN INFOGRAPHICS DESIGNER

The growth in importance of metrics, KPIs, dashboards, and business intelligence applications has been spectacular. Unfortunately, the result has often been information overload, primarily with dashboard reporting systems. Today, we tend to add in more artwork than we need, a trend that resulted in a new term, infographics. Some problems with the growth of infographics are:

- There is a heavy focus on designs, colors, images, and text rather than the quality of the information being presented.
- A decline in the quality of the information makes it difficult for stakeholders to use the data properly.
- There are too many pretty graphics that can be misleading and hard to understand.
- The dashboard has been converted from a project management performance tool to a marketing/sales tool.
- Some graphic artists do not understand or utilize information visualization best practices.
- Some dashboard designers use lots of glitter, which prevents clear communications and focuses on perhaps the wrong objectives.
- Some designers simply do not understand how long it takes to properly design a project dashboard.

What should be the criteria for selecting the right infographic designer?

Copyright © Scott Maxwell/Fotolia

We must have better and clearer representation of the metrics we select and how they will be used. Not all graphic designers are qualified to work on project dashboards. Typical selection criteria for a project dashboard designer might be an understanding of:

- The basics of project management
- The dashboard audience
- The purpose of the data
- How the data will be used by the viewers

- How the data will be updated and how often
- Security surrounding display of the data

Dashboard success is achieved when the status of the project is quite apparent from the dashboard metrics. The dashboard designer must understand that:

- The visualization of data must appear in a manner that is easily understood.
- The primary focus is for the data needed for informed decision making rather than the means of display.
- Dashboard viewers must be able to quickly identify those items that need immediate attention.
- If dashboard viewers must scroll around the dashboard or constantly change screens, then the overall purpose of having a dashboard has not been achieved.

Regardless of how good the dashboard designers work, dashboard designers must understand that first-time viewers of a dashboard may misinterpret what they are seeing and may draw the wrong conclusions. Some companies set up dashboard pilot runs to verify that the stakeholders:

- Understand what they are viewing
- Arrive at the correct conclusions
- Have faith in the dashboard concepts
- Are willing to use the information for decision making

5.11 PROJECT HEALTH CHECK METRICS

Projects seem to progress quickly until they are about 60–70% complete. During that time, everyone applauds that work is progressing as planned. Then, perhaps without warning, the truth comes out, possibly due to significant scope creep, and we discover that the project is in trouble. This occurs because of:

- Our disbelief in the value of using the project's metrics correctly
- Selecting the wrong metrics
- Our fear of what project health checks may reveal

How do we determine a project's health?

Some project managers have an incredible fixation with project metrics and numbers, believing that metrics are the Holy Grail in determining status. This holds true for both PM 1.0 and PM 2.0. Most projects in PM 1.0 seem to focus on only two metrics: time and cost. These are the primary metrics in all EVMSs. While these two metrics "may" give you a reasonable representation where you are today, using these two metrics to provide forecasts into the future are "grey" areas and may not indicate future problem areas that could prevent a successful and timely completion of the project. At the other end of the spectrum we have managers who have no faith in the metrics and therefore focus on vision, strategy, leadership, and prayers. Rather than relying on metrics alone, the simplest solution might be to perform periodic health checks on the project. In doing this, four critical questions must be addressed:

- Who will perform the health check?
- Will the interviewees be honest in their responses?

- Will we need to look at more metrics than currently being used?
- Will management and stakeholders overreact to the truth?

The surfacing of previously unknown or hidden issues could lead to loss of employment, demotions, or project cancellation. Yet project health checks offer the greatest opportunity for early corrective action to save a potentially failing project. Health checks can also discover future opportunities.

People tend to use audits and health checks synonymously. Both are designed to ensure successful, repeatable project outcomes, and both must be performed on projects that appear to be heading for a successful outcome as well as those that seem destined to fail. There are lessons learned and best practices that can be discovered from both successes and failures. Also, detailed analysis of a project that appears to be successful at the moment might bring to the surface issues that show that the project is really in trouble.

Just because a project is on time and/or within the allotted budget does not guarantee success. The end result could be that the deliverable has poor quality such that it is unacceptable to the customer. In addition to time and cost, project health checks focus on quality, resources, benefits, and requirements, just to name a few. The true measure of the project's future success is the value that the customers see at the completion of the project. Health checks must therefore be value focused. Audits, on the other hand, usually do not focus on value.

Health checks can function as an ongoing tool by being performed randomly when needed or periodically throughout various life-cycle stages. However, there are specific circumstances that indicate that a health check should be accomplished quickly. These include:

- Significant scope creep
- Escalating costs accompanied by a deterioration in value and benefits
- Schedule slippages that cannot be corrected
- Missed deadlines for deliverables
- Poor morale accompanied by changes in key project personnel

Periodic health checks, if done correctly, eliminate ambiguity such that true status can be determined. The benefits of health checks include:

- Determining the current status of the project
- Identifying problems early enough such that sufficient time exists for corrective action to be taken
- Identifying the CSFs that will support a successful outcome or the critical issues that can prevent successful delivery
- Identifying lessons learned, best practices, and CSFs that can be used on future projects
- Evaluating compliance to and improvements for the EPM methodology
- Identifying which activities may require or benefit from additional resources
- Identifying present and future risks as well as possible risk mitigation strategies
- Determining if the benefits and value will be there at completion
- Determining if euthanasia is required to put the project out of its misery
- The development of or recommendations for a fix-it plan

There are misconceptions about project health checks. Some of these are:

- The person doing the health check does not understand the project or the corporate culture, thus wasting time.
- The health check is too costly for the value we will get by performing it.
- The health check ties up critical resources in interviews.
- By the time we get the results from the health check, either it is too late to make changes or the nature of the project may have changed.

One of the challenges facing companies is whether the health check should be conducted by internal personnel or by external consultants. The risk with using internal personnel is that they may have loyalties or relationships with people on the project team and therefore may not be totally honest in determining the true status of the project or in deciding who was at fault. Using external consultants or facilitators is often the better choice. External facilitators can bring to the table:

- A multitude of forms, guidelines, templates, and checklists used in other companies and similar projects
- A promise of impartiality and confidentiality
- A focus on only the facts and hopefully be free of politics
- An environment where people can speak freely and vent their personal feelings
- An environment that is relatively free from other day-to-day issues

There are three life-cycle phases for project health checks:

- Review of the business case and the project's history
- Research and discovery of the facts
- Preparation of the health check report

Reviewing the business case and project's history may require the health check leader to have access to proprietary knowledge and financial information. The leader may have to sign nondisclosure agreements and also noncompeting clauses before being allowed to perform the health check.

In the research-and-discovery phase, the leader prepares a list of questions that need to be answered. The list can be prepared from the **PMBOK® Guide**'s domain areas or areas of knowledge. The questions can also come from the knowledge repository in the consultant's company and may appear in the form of templates, guidelines, checklists, or forms. The questions can change from project to project and industry to industry.

Some of the critical areas that must be investigated are:

- Performance against baselines
- Ability to meet forecasts
- Benefits and value analyses
- Governance
- Stakeholder involvement
- Risk mitigation
- Contingency planning

If the health check requires one-on-one interviews, the health check leader must be able to extract the truth from interviewees that have different interpretations or conclusions about the status of the project. Some people will be truthful whereas others will either say what they believe the interviewer wants to hear or distort the truth as a means of self-protection.

The final phase is the preparation of the report. This should include:

- A listing of the issues
- Root-cause analyses, possibly including identification of individuals that created the problems
- Gap analysis
- Opportunities for corrective action
- A get-well or fix-it plan

Project health checks are not "Big Brother Is Watching You" activities. Rather, they are part of project oversight. Without these health checks, the chances for project failure are significantly increased. Project health checks also provide us with insight on how to keep risks under control. Performing health checks and taking corrective action early are certainly better than having to manage a distressed project.

With PM 2.0, we are trying to determine a set of core metrics that can be used to measure the health of the project as it progresses. The six core metrics currently used appear in Table 5-4.

However, based upon the complexity of the project, these six core metrics may not be enough. We may need additional metrics, such as:

- Deliverables (in progress): late versus on time
- Deliverables (completed): accepted versus rejected
- Management reserve: amount available versus used
- Risks: number of risks in each core metric category
- Action items: number of action items in each core category
- Action items aging: the number of action items that are over 1 month, 2 months, 3 or more months late

One discrete metric does not often provide a clear picture of the health of a project. Continuous data are needed to illustrate the trend or identify the real problems.

TABLE 5-4 Core Metrics

Metric	Measurement
Time	Schedule performance index
Cost	Cost performance index
Resources	Quality or pay grade and actual number versus planned staff
Scope (or scope creep)	Number of scope change requests approved, denied, and pending
Quality	Number of defects against a user acceptance criteria
Action items	Number of action items still open

5.12 MAINTAINING PROJECT'S DIRECTION

Metrics can show us that the project is headed in the agreed-upon direction that was established at the onset of the project. Unfortunately, the agreement may have been reached just to get the project started. After go-ahead, some stakeholders may try to change the direction. Senior managers, clients, and stakeholders in general may have hidden agendas concerning the project. Even though they may agree on the direction initially, they may try to change the project's direction once the project begins. Not all stakeholders may want the project to succeed. Some stakeholders may see the successful completion of the project as a reduction in the size of their empire, a loss of power and authority in the future, or having a negative impact on the size of their yearly bonuses.

Will stakeholders always agree on the project's direction?

Copyright © Scott Maxwell/ LuMaxArt/Shutterstock

As projects become larger and more complex, there is a much greater risk of scope changes that can result in modifications to the scope, time, and cost baselines. People with decision-making power may see this as a window of opportunity to change the direction of the project for their own personal desires. There are several other factors that can make it difficult to maintain the project's direction. Some of them are:

- Changes in the enterprise environmental factors
- High turnover in governance committee members
- Personal desires of the most influential financial stakeholder
- A threat by the most influential stakeholder to withdraw funding
- Political instability in the host country
- Elections resulting in new leadership that is not in agreement with the direction of the project
- Political intervention
- Impact of a labor shortage and union negotiations
- Financial solvency of some critical suppliers
- Changes in procurement policies
- Changes in government legislation
- Not having the right skilled labor for the changes

Decision makers often do not understand the impact that changes can have on a project. The ideal situation for PM 2.0 would be a short course for all governance stakeholders. However, this is probably wishful thinking to get them to attend.

5.13 METRICS AND VIRTUAL TEAMS

Historically, project management was a face-to-face environment where team meetings involved all players convening together in one room. The team itself may even be colocated. Today, because of the size and complexity of projects, it is impossible to find all team members located under one roof. The team members can be located anywhere in the world and the project is then managed using virtual teams.

With virtual teams, the project manager may never meet many of the team members face to face and must then rely upon effective communication skills to create a cohesive team. Communication management may very well be the most important skill

How important are metrics when using virtual teams?

when using virtual teams. Metrics then become extremely important as a communication tool.

Culture and technology can have a major impact on the performance of virtual teams. The importance of culture cannot be understated. Duarte and Snyder identify four important points to remember concerning the impact of culture on virtual teams[11]:

- There are national cultures, organizational cultures, functional cultures, and team cultures. They can be sources of competitive advantages for virtual teams that know how to use cultural differences to create synergy. Team leaders and members who understand and are sensitive to cultural differences can create more robust outcomes than can members of homogeneous teams with members who think and act alike. Cultural differences can create distinctive advantages for teams if they are understood and used in positive ways.
- The most important aspect of understanding and working with cultural differences is to create a team culture in which problems can be surfaced and differences can be discussed in a productive, respectful manner.
- It is essential to distinguish between problems that result from cultural differences and problems that are performance based.
- Business practices and business ethics vary in different parts of the world. Virtual teams need to clearly articulate approaches to these that every member understands and abides by.

When working with a colocated team or people we see face to face, we tend to neglect the impact of culture on the metrics we select and the design of the metrics dashboard. However, with virtual teams cultural considerations become important. Certain colors may have political or religious ramifications. We also may not know if they are interpreting the data correctly. Sometimes, we do not know if anyone other than the virtual team members will be reviewing the metric data. Without this knowledge, care must be taken as to the inclusion of proprietary information on the dashboard.

5.14 METRIC MANIA

Metric mania is like a disease. It is characterized as someone who is so enamored with metrics that they try to create as many metrics as they can even though most of the metrics may have no practical or informational value. Another form of metric mania occurs with infographic designers who are infatuated more with various colors, images, and glitter than with the quality of the information that needs to be represented.

Perhaps the best way of controlling metric mania is through the PMO. Some companies are considering having one person in the PMO with the title of chief

---

[11] D. L. Duarte and N. T. Snyder, *Managing Virtual Teams*, Jossey-Bass, San Francisco, CA, an imprint of John Wiley & Sons, New York, 2001, p.70. Reprinted by permission of John Wiley & Sons.

performance officer. This person would control the quality and value of all metrics used and whether or not they should be part of the permanent metrics library.

The hardest form of metric mania to control is stakeholder or client metric mania. As we stated earlier in this chapter, companies are creating metric libraries. At the beginning of a project, we work with the stakeholders to define project success and then look at the metrics in the library that will be of value in measuring success. Stakeholder metric mania occurs when the stakeholders want all of the metrics in the library displayed on several dashboard screens even if it requires the use of several dashboards for each stakeholder. Other than the cost of maintaining multiple dashboards, information overload will occur to the point where stakeholders may not be able to identify what information is critical for informed decision making.

Can we prevent "metric mania"?

5.15 METRIC TRAINING SESSIONS

Regardless of how knowledgeable the organization is in project management, training programs on metrics management should take place. Without training, people cannot differentiate between a good and bad metric and may end up learning from their own mistakes rather than the mistakes of others. The organization may end up adopting bad metrics that provide meaningless information and alienate the stakeholders such that support for metrics management may be derailed. Most people believe that they have a good understanding of metrics when, in fact, the reverse is often true.

What education will be required for effective project management?

Effective metric management training programs do more than teach people about metrics. They also discuss the importance of cultivating a metrics culture, beginning with identifying the need for senior management support. Items included in the beginning of the training might be:

- Some executives will not take ownership for a metrics management system for fear of looking bad in the eyes of their colleagues if the metrics reporting system is not accepted by the workers or fails to provide meaningful results. Obtaining their support is essential.
- Executives will not support a metrics management system that looks like pay for performance for executives and can affect their bonuses and chances for promotion. Executives may then encourage the use of only those metrics that make them look good.
- There must be an institutional belief in the value of a metrics management program.
- The belief must be visibly supported by senior management.
- The metrics must be used by executives and managers for informed decision making.
- The metrics must be aligned with both corporate objectives and project objectives.
- People must be open and receptive to change.

- The organization must recognize the improved performance.
- The organization must be willing to support the identification, collection, measurement, and reporting of metrics.
- The organization must understand that effective metrics management is based upon employee buy-in rather than from purchasing software and hiring consultants.
- Executives must make sure that they do not inadvertently undermine metric management implementation.
- Executives must be willing to see or change bad habits that can be corrected using metrics management.
- Metric management implementation must focus on what needs to be improved the most.
- Metric management implementation can fail if the data are not reliable.
- However, expecting 100% reliability may be unrealistic because of human error.

The next part of the training program emphasizes the core concepts behind identifying, measuring, tracking, and reporting metric information. Participants must understand that:

- Identifying metrics will require "data mining" and the information discovered must be displayed in an easy-to-understand format.
- The purpose of metric management is not only to provide information but also to encourage the discovery of continuous improvement opportunities.

People must understand that it may be impossible to create a perfect set of metrics for each project. Mistakes will be made initially. It is more important to get buy-in from everyone, including stakeholders, than to look for perfection.

5.16 METRIC OWNERS

Complacency in metrics management can lead to bad results. Once a project is completed, workers tend to forget about the metrics they used and whether or not the metrics can be improved upon. The result is a lack of continuous improvements with metrics. Differentiation between good and bad metrics is not easy. People may not know that they are using the wrong metric or that better metric measurement techniques exist.

Companies today are creating a position called a metric owner. Most of the time, the metric owner is someone who has attended the metrics management courses and volunteers to serve as a metric owner. This is a person who takes the responsibility for continuous improvement of a given metric by looking at better ways to measure the metric, report performance, and display the metric on a project dashboard. The corporate metric owner:

- Understands the company's culture
- Has the respect of the labor force
- Can foster the support for metric management
- Knows how to overcome resistance

Can we differentiate between good and bad metrics?

Metric owners are usually functional employees who report with a dotted line to the PMO for metric updates and participate in the periodic reevaluation of metrics.

5.17 ANSWERING METRIC QUESTIONS

Some people have the mistaken belief that an effective metrics management program will reduce the pile of questions that the project team must address. In the long term, there may be fewer questions. But in the near term, there can be significantly more questions because:

Will metrics management require that we answer piles of questions?

- Projects are becoming more complex and more information is needed.
- Project governance is now committee governance and each committee member may have different information needs.
- Not all of the governance committee members will understand metrics management.
- Not all governance committee members will have an understanding of project management.
- Governance committee members may bring with them hidden agendas which require answers to questions that cannot come from metric information.
- The project team must validate that all first-time viewers of dashboards understand what they are seeing.
- As projects get longer in duration, the risk of changes in the enterprise environmental factors, changes in the assumptions, and the identification of new constraints will lead to more questions.
- Changing from PM 1.0 to PM 2.0 will most certainly create additional questions.
- Internal users of the new metrics management system will have additional questions.

As people become familiar with PM 2.0 and use many more metrics, the number of questions will most likely decrease. But in the short term perhaps more than 90% of the project manager's time will be spent in communications.

DISCUSSION QUESTIONS

The discussion questions are for classroom use to stimulate group thinking about PM 2.0. There are no right or wrong answers to most of the questions.

1. Is there a difference between enterprise resource planning and project staffing?
2. What are the differences between metrics and KPIs?
3. Can the definition of a KPI change from project to project?
4. Who makes the final decision as to what metrics will be displayed on dashboards?
5. Who should be assigned as a metric owner?
6. Are health checks performed using metrics or KPIs?

7. Who should be the prime source for answering questions regarding project metrics?
8. Should communications with virtual teams be accomplished with metrics, KPIs, or both?
9. Should both metrics and KPIs be regarded as business intelligence information?
10. Can metric mania be controlled? If so, how?

CHAPTER 6

PROJECT MANAGEMENT METHODOLOGIES: 1.0 VERSUS 2.0

6.0 INTRODUCTION

The importance of a good methodology cannot be understated. It will not only improve your performance during project execution but also allow for better customer relations and build customer confidence. Good methodologies are a necessity for PM 2.0 and can also lead to sole-source or single-source procurement contracts.

Creating a workable methodology for project management is no easy task. One of the biggest mistakes made is developing a different methodology for each type of project. Another is failing to integrate the project management methodology and project management tools into a single process, if possible. When companies develop project management methodologies and tools in tandem, four benefits emerge. First, the work may be accomplished with fewer scope changes. Second, the processes are designed to create minimal disturbance to ongoing business operations. Third, if metrics are used effectively, the methodology will produce repeatable effective and efficient results. Fourth, methodologies have exit gates and off ramps whereby poorly performing projects can be canceled or redirected.

The rigidity of PM 1.0 methodologies made it difficult for some project managers to customize the methodology to a particular client. With PM 2.0, the project managers have significantly more freedom in how to apply the methodology. The flexibility should not interfere with the intent of the methodology to provide guidance in how to meet the project's requirements.

6.1 PM 2.0 DEFINITION OF PROJECT MANAGEMENT EXCELLENCE

Excellence in project management is often regarded as a continuous stream of successfully managed projects. Without a project management methodology, repetitive, successfully completed projects may be difficult to achieve.

Today, everyone seems to agree somewhat on the necessity for a project management methodology. However, there is still disagreement on the definition of excellence in project management, the same way that companies have different definitions for project success.

In earlier chapters, we defined project success as creating business value within the competing constraints. PM 2.0 project management excellence can then be defined as having a stream of successfully managed projects that are completed within the competing constraints and satisfy the business value creation requirements of all stakeholders whether they are internal or external stakeholders.

6.2 NEED FOR A METHODOLOGY

Simply having a project management methodology and following it will not lead to success and excellence in project management. The need for continuous improvements in the system may be crucial. External factors can have a strong influence on the success or failure of a company's project management methodology. Change is a given in the current business climate, and there is no sign that the future will be any different. The rapid changes in technology that have driven changes in project management over the past two decades are not likely to subside. Another trend, the increasing sophistication of consumers and clients, is likely to continue, not go away. Cost and quality control have become virtually the same issue in many industries. Other external factors include rapid mergers and acquisitions and the need for real-time communications.

Project management methodologies are organic processes and need to change as the organization changes in response to the ever-evolving business climate. Such changes, however, require that managers on all levels be committed to the changes and develop a vision that calls for the development and use of project management systems along with the rest of the organization's other business systems.

Today, companies are managing their business by projects. This is true for both non-project-driven and project-driven organizations. Virtually all activities in an organization can be treated as some sort of project. Therefore, it is only fitting that well-managed companies regard a project management methodology as a way to manage the entire business rather than just projects. Business processes and project management processes will be merged together as the project manager is viewed as the manager of part of a business rather than just the manager of a project.

Developing a standard project management methodology is not for every company. For companies with small or short-term projects, such formal systems may not be cost effective or appropriate. However, for companies with large or ongoing projects, developing a workable project management system is mandatory.

For example, a company that manufactures home fixtures had several project development protocols in place. When they decided to begin using project management systematically, the complexity of the company's current methods became apparent. The company had multiple system development methodologies based on the type of project. This became awkward for employees who had to struggle with a different methodology for each project. The company then opted to create a general, all-purpose methodology for all projects. The new methodology had flexibility built into it. According to one spokesman for the company:

Our project management approach, by design, is not linked to a specific systems development methodology. Because we believe that it is better to use a (standard) systems development methodology than to decide which one to use, we have begun development of a guideline systems development methodology specific for our organization. We have now developed prerequisites for project success. These include:

A well-patterned methodology

A clear set of objectives

Well-understood expectations

Thorough problem definition

During the late 1980s, merger mania hit the banking community. With the lowering of costs due to economies of scale and the resulting increased competitiveness, the banking community recognized the importance of using project management for mergers and acquisitions. The quicker the combined cultures became one, the less the impact on the corporation's bottom line.

The need for a good methodology became apparent, according to a spokesperson at one bank:

The intent of this methodology is to make the process of managing projects more effective: from proposal to prioritization to approval through implementation. This methodology is not tailored to specific types or classifications of projects, such as system development efforts or hardware installations. Instead, it is a commonsense approach to assist in prioritizing and implementing successful efforts of any jurisdiction.

In 1996, the information services (IS) division of one bank formed an IS reengineering team to focus on developing and deploying processes and tools associated with project management and system development. The mission of the IS reengineering team was to improve performance of IS projects, resulting in increased productivity, cycle time, quality, and satisfaction of the projects' customers.

According to a spokesperson at the bank, the process began as follows:

Information from both current and previous methodologies used by the bank was reviewed, and the best practices of all these previous efforts were incorporated into this document. Regardless of the source, project methodology phases are somewhat standard fare. All projects follow the same steps, with the complexity, size, and type of project dictating to what extent the methodology must be followed. What this methodology emphasizes are project controls and the tie of deliverables and controls to accomplishing the goals.

To determine the weaknesses associated with past project management methodologies, the IS reengineering team conducted various focus groups. These focus groups concluded that there was:

- Lack of management commitment
- Lack of a feedback mechanism for project managers to determine the updates and revisions needed to the methodology
- Lack of adaptable methodologies for the organization
- Lack of training curriculum for project managers on the methodology
- Lack of focus on consistent and periodic communication on the methodology deployment progress

- Lack of focus on the project management tools and techniques
- Based on this feedback, the IS reengineering team successfully developed and deployed a project management and system development methodology. The bank did an outstanding job of creating a methodology that reflects guidelines rather than policies and provides procedures that can easily be adapted on any project in the bank.
- While these examples are from a PM 1.0 environment, we still recognize the need to continue to do many of these with PM 2.0.

6.3 NEED FOR AN ENTERPRISEWIDE METHODOLOGY

Most companies today seem to recognize the need for one or more project management methodologies but either create the wrong methodologies or misuse the methodologies that have been created. Many times, companies rush into the development or purchasing of a methodology without any understanding of the need for one other than the fact that their competitors have a methodology. Jason Charvat states[1]:

> Using project management methodologies is a business strategy allowing companies to maximize the project's value to the organization. The methodologies must evolve and be "tweaked" to accommodate a company's changing focus or direction. It is almost a mind-set, a way that reshapes entire organizational processes: sales and marketing, product design, planning, deployment, recruitment, finance, and operations support. It presents a radical cultural shift for many organizations. As industries and companies change, so must their methodologies. If not, they're losing the point.

Methodologies with PM 1.0, such as the example with the bank in the previous section, were often based upon rigid policies and procedures. Project managers believed they were wearing handcuffs and were not allowed to deviate from the policies and procedures in the methodology. Simply stated, executives did not trust the project managers to make the right decisions. Rigid methodologies were used as a means of executive control. Methodologies with PM 2.0 are a set of forms, guidelines, templates, and checklists that can be applied to a specific project or situation. Customers want the contractor to use a methodology that is flexible enough such that it can be adapted to the customer's business model. If the methodology does not possess this flexibility, then it may not be possible to maximize the value that the customer expects. PM 2.0 methodologies remove the handcuffs from the project managers and allow them the flexibility needed to select those forms, guidelines, templates, and checklists that are directly applicable to the customer's business model and needs.

Even with the addition of flexibility, it may not be possible to create a single enterprisewide methodology that can be applied to each and every project. Some companies have been successful doing this, but there are still many companies that successfully maintain more than one methodology. Unless the project manager is capable of tailoring the EPM methodology to his or her needs, more than one methodology may be necessary.

There are several reasons why good intentions often go astray. At the executive levels, methodologies can fail if the executives have a poor understanding of what a methodology is and believe that a methodology is:

---

[1] J. Charvat, *Project Management Methodologies*, Wiley, Hoboken, NJ, 2003, p. 2.

- A quick fix
- A silver bullet
- A temporary solution
- A cookbook approach for project success[2]

At the working levels, methodologies can also fail if they:

- Are abstract and high level
- Contain insufficient narratives to support these methodologies
- Are not functional or do not address crucial areas
- Ignore the industry standards and best practices
- Look impressive but lack real integration into the business
- Use nonstandard project conventions and terminology
- Compete for similar resources without addressing this problem
- Do not have any performance metrics
- Take too long to complete because of bureaucracy and administration[3]

Deciding on the type of methodology is not an easy task. There are many factors to consider, such as:

- The overall company strategy—how competitive are we as a company?
- The size of the project team and/or scope to be managed
- The priority of the project
- How critical the project is to the company
- How flexible the methodology and its components are[4]

Project management methodologies are created around the project management maturity level of the company and the corporate culture. If the company is reasonably mature in project management and has a culture that fosters cooperation, effective communications, teamwork, and especially trust, then a highly flexible methodology can be created based upon guidelines, forms, checklists, and templates. Trust is a prime characteristic of PM 2.0. Project managers can pick and choose the parts of the methodology that are appropriate for a particular client. Organizations that do not possess the above characteristics rely heavily upon methodologies constructed with rigid policies and procedures, thus creating significant paperwork requirements with accompanying cost increases and removing the flexibility that the project manager needs for adapting the methodology to the needs of a specific client.

Jason Charvat describes two types of methodologies as light methodologies and heavy methodologies.[5]

Light Methodologies

Ever-increasing technological complexities, project delays, and changing client requirements brought about a small revolution in the world of development methodologies.

---

[2] Ibid., p. 4.
[3] Ibid., p. 5.
[4] Ibid., p. 66.
[5] Ibid., pp. 102–104.

A totally new breed of methodology—which is agile and adaptive and involves the client every part of the way—is starting to emerge. Many of the heavyweight methodologists were resistant to the introduction of these "lightweight" or "agile" methodologies (Fowler, 2001[6]). These methodologies use an informal communication style. Unlike heavyweight methodologies, lightweight projects have only a few rules, practices, and documents. Projects are designed and built on face-to-face discussions, meetings, and the flow of information to the clients. The immediate difference of using light methodologies is that they are much less documentation oriented, usually emphasizing a smaller amount of documentation for the project.

Heavy Methodologies

The traditional project management methodologies [i.e., systems development life-cycle (SDLC) approach] are considered bureaucratic or "predictive" in nature and have resulted in many unsuccessful projects. These heavy methodologies are becoming less popular. These methodologies are so laborious that the whole pace of design, development, and deployment slows down—and nothing gets done. Project managers tend to predict every milestone because they want to foresee every technical detail (i.e., software code or engineering detail). This leads managers to start demanding many types of specifications, plans, reports, checkpoints, and schedules. Heavy methodologies attempt to plan a large part of a project in great detail over a long span of time. This works well until things start changing, and the project managers inherently try to resist change.

Enterprise project management methodologies can enhance the project planning process as well as providing some degree of standardization and consistency. Companies have come to the realization that EPM methodologies work best if the methodology is based upon templates rather than rigid policies and procedures. The International Institute for Learning has created a Unified Project Management Methodology (UPMM™) with templates categorized according to the **PMBOK® Guide, 5th edition,** areas of knowledge[7]:

Communication:
> Project Charter
> Project Procedures Document
> Project Change Requests Log
> Project Status Report
> PM Quality Assurance Report
> Procurement Management Summary
> Project Issues Log
> Project Management Plan
> Project Performance Report

Cost:
> Project Schedule
> Risk Response Plan and Register
> Work Breakdown Structure (WBS)
> Work Package

---

[6] M. Fowler, *The New Methodology, Thought Works*, 2001, available: www.martinfowler.com/articles.
[7] Unified Project Management Methodology (UPMM™) is registered, copyrighted, and owned by the International Institute for Learning, Inc. © 2014. Reproduced by permission.

Cost Estimates Document
Project Budget
Project Budget Checklist

Human Resources:
Project Charter
Work Breakdown Structure (WBS)
Communications Management Plan
Project Organization Chart
Project Team Directory
Responsibility Assignment Matrix (RAM)
Project Management Plan
Project Procedures Document
Kick-off Meeting Checklist
Project Team Performance Assessment
Project Manager Performance Assessment

Integration:
Project Procedures Overview
Project Proposal
Communications Management Plan
Procurement Plan
Project Budget
Project Procedures Document
Project Schedule
Responsibility Assignment Matrix (RAM)
Risk Response Plan and Register
Scope Statement
Work Breakdown Structure (WBS)
Project Management Plan
Project Change Requests Log
Project Issues Log
Project Management Plan Changes Log
Project Performance Report
Lessons Learned Document
Project Performance Feedback
Product Acceptance Document
Project Charter
Closing Process Assessment Checklist
Project Archives Report

Procurement:
Project Charter
Scope Statement
Work Breakdown Structure (WBS)
Procurement Plan
Procurement Planning Checklist
Procurement Statement of Work (SOW)

Request for Proposal Document Outline
Project Change Requests Log
Contract Formation Checklist
Procurement Management Summary

Quality:
Project Charter
Project Procedures Overview
Work Quality Plan
Project Management Plan
Work Breakdown Structure (WBS)
PM Quality Assurance Report
Lessons Learned Document
Project Performance Feedback
Project Team Performance Assessment
PM Process Improvement Document

Risk:
Procurement Plan
Project Charter
Project Procedures Document
Work Breakdown Structure (WBS)
Risk Response Plan and Register

Scope:
Project Scope Statement
Work Breakdown Structure (WBS)
Work Package
Project Charter

Time:
Activity Duration Estimating Worksheet
Cost Estimates Document
Risk Response Plan and Register Medium
Work Breakdown Structure (WBS)
Work Package
Project Schedule
Project Schedule Review Checklist

Stakeholder Management:
Stakeholder Register
Stakeholder Management Plan
Power/Interest Grid
Stakeholders Engagement Assessment Matrix

6.4 BENEFITS OF A STANDARDIZED METHODOLOGY

For companies that understand the importance of a standard methodology that contains some degree of flexibility, the benefits are numerous. These benefits can be classified as both short- and long-term benefits. Short-term benefits were described by one company as:

- Decreased cycle time and lower costs
- Realistic plans with greater possibilities of meeting all constraints
- Better communications as to "what" is expected from groups and "when"
- Greater customer satisfaction
- Greater likelihood of follow-on work
- Greater likelihood of creating business value
- Feedback: lessons learned

These short-term benefits focus on metrics/KPIs or, simply stated, the execution of project management. Long-term benefits seem to focus more upon CSFs, business value created, and customer satisfaction. Long-term benefits of development and execution of a world-class methodology include:

- Faster "time to market" through better scope control
- Lower overall program risk
- Better risk management, which leads to better decision making
- Greater customer satisfaction and trust, which lead to increased business and expanded responsibilities for the tier 1 suppliers
- Emphasis on customer satisfaction and value-added rather than internal competition between functional groups
- Customer treating the contractor as a "partner" rather than as a commodity
- Contractor assisting the customer during strategic planning activities

Perhaps the largest benefit of a world-class methodology is the acceptance and recognition by your customers. If one of your critically important customers develops its own methodology, that customer could "force" you to accept it and use it in order to remain a supplier. But if you can show that your methodology is superior or equal to the customer's, your methodology will be accepted, and an atmosphere of trust will prevail.

One contractor recently found that its customer had so much faith in and respect for its methodology that the contractor was invited to participate in the customer's strategic planning activities. The contractor found itself treated as a partner rather than as a commodity or just another supplier. This resulted in sole-source procurement contracts for the contractor.

Developing a somewhat flexible standard methodology that encompasses the majority of a company's projects and is accepted by the entire organization is a difficult undertaking. The hardest part might very well be making sure that the methodology supports both the corporate culture and the goals and objectives set forth by management. Methodologies that require changes to a corporate culture may not be well accepted by the organization. Nonsupportive cultures can destroy even seemingly good project management methodologies.

During the 1980s and 1990s, several consulting companies developed their own project management methodologies, most frequently for information systems projects, and then pressured their clients into purchasing the methodology rather than helping their clients develop a methodology more suited to the clients' needs. Although there may have been some successes, there appeared to be significantly more failures than successes. A hospital purchased a $130,000 project management methodology with the belief that this would be the solution to its information system needs. Unfortunately, senior management made the purchasing decision without consulting the workers who would be using the system. In the end, the package was never used.

Another company purchased a similar package, discovering too late that the package was inflexible and the organization, specifically the corporate culture, would need to change to use the project management methodology effectively. The vendor later admitted that the best results would occur if no changes were made to the methodology.

These types of methodologies are extremely rigid and based on policies and procedures. The ability to custom design the methodology to specific projects and cultures was nonexistent, and eventually these methodologies fell by the wayside—but after the vendors made significant profits. Good methodologies must be flexible. The question, of course, is how much flexibility can it withstand before problems occur?

6.5 CRITICAL COMPONENTS

It is almost impossible to become a world-class company with regard to project management without having a world-class methodology. Years ago, perhaps only a few companies really had world-class methodologies. Today, because of the need for survival and stiffening competition, there are numerous companies with good methodologies.

The characteristics of a world-class methodology include:

- Maximum of six life-cycle phases
- Life-cycle phases overlap
- End-of-phase gate reviews
- Integration with other processes
- Continuous improvement (i.e., hear the voice of the customer)
- Customer oriented (interface with customer's methodology)
- Companywide acceptance
- Use of templates (level 3 WBS)
- Critical path scheduling (level 3 WBS)
- Effective use of a dashboard reporting system
- Minimization of paperwork

Generally speaking, each life-cycle phase of a project management methodology requires paperwork, control points, and perhaps special administrative requirements. Having too few life-cycle phases is an invitation for disaster, and having too many life-cycle phases may drive up administrative and control costs. Most companies prefer a maximum of six life-cycle phases. The control points are an absolute necessity to assure that the project should continue. When methodologies based upon PM 2.0 requirements have flexibility as stated previously, the project manager's control (or decision-making) points must be aligned with the customer's required control points.

Historically, life-cycle phases were sequential in nature. However, because of the necessity for schedule compression, life-cycle phases today will overlap. The amount of overlap will be dependent upon the magnitude of the risks the project manager will take. The more the overlap, the greater the risk. Mistakes made during overlapping activities are usually more costly to correct than mistakes during sequential activities. Overlapping life-cycle phases requires excellent up-front planning.

End-of-phase gate reviews are critical for control purposes and verification of interim milestones. With overlapping life-cycle phases, there are still gate reviews at the end of each phase, but they are supported by intermediate reviews during the life-cycle phases.

World-class project management methodologies are integrated with other management processes such as change management, risk management, total quality management, and concurrent engineering. This produces a synergistic effect which minimizes paperwork, minimizes the total number of resources committed to the project, and allows the organization to perform capacity planning to determine the maximum workload that the organization can endure.

World-class methodologies are continuously enhanced through metric/KPI reviews, lessons-learned updates, benchmarking, and customer recommendations. The methodology itself could become the channel of communication between the customer and contractor. Effective methodologies foster customer trust and minimize customer interference in the project.

Project management methodologies must be easy for workers to use as well as covering most of the situations that can arise on a project. Perhaps the best way is to have the methodology placed in a manual that is user friendly.

Excellent methodologies try to make it easier to plan and schedule projects. This is accomplished by using templates for the top three levels of the WBS. Simply stated, using WBS level 3 templates, standardized reporting with standardized terminology exists. The differences between projects will appear at the lower levels (i.e., levels 4–6) of the WBS. This also leads to a minimization of paperwork.

Today, companies seem to be promoting the use of the project charter concept as a component of a methodology, but not all companies create the project charter at the same point in the project life cycle. There are three locations where the charter can be prepared:

- The charter is prepared immediately after the feasibility study is completed. At this point, the charter contains the results of the feasibility study as well as documentation of any assumptions and constraints that were considered. This could be part of the project portfolio creation activities. The charter is then revisited and updated once this project is selected and work begins.
- In the second choice, which seems to be the preferred method, the charter is prepared after the project is selected and the project manager has been assigned. The charter includes the authority granted to the project manager, but for this project only.
- In the third method, the charter is prepared after detail planning is completed. The charter contains the detailed plan that may have been developed during the portfolio selection process. Management will not sign the charter until after detail planning is approved by senior management. Then, and only then, does the company officially sanction the project. Once management signs the charter, the charter becomes a legal agreement between the project manager and all involved line managers as to what deliverables will be met and when.

6.6 FROM METHODOLOGIES TO FRAMEWORK

Over the years, companies have come to accept the **PMBOK® Guide** as the gospel of project management. But we must remember that the **PMBOK® Guide** is still just a guide and not necessarily the actual body of knowledge. Creating a perfect body of knowledge would have to be based upon the size, nature, and complexity of a firm's projects, the type of industry in which they compete, and the percentage of work within the firm that must utilize project management. Some project management educators argue that there are three ways to manage projects: the right way, the wrong way, and the **PMBOK® Guide** way. Although companies create project management methodologies based upon the **PMBOK® Guide**, the methodologies rarely follow it exactly. A large percentage of the areas of knowledge may be used, but the focus could be more heavily aligned to the domain areas rather than the areas of knowledge. Processes may be more important than knowledge areas. Processes focus heavily upon frameworks that allow the project managers the flexibility needed for PM 2.0.

There is certainly a need for software in project management. But the critical question is as follows: Should we select a project management software tool and then design our project management methodology around that tool or should we design our methodology first and then select the appropriate tool to fit our methodology? The latter is certainly the better choice. Software is simply a tool. Projects are managed by people, not tools. People manage tools; tools do not manage people. A proper understanding of software and its capability is essential at the onset of project management implementation but should not be viewed as a replacement for good processes.

While project management methodologies seem to be in favor today, there is a growing trend toward replacing methodologies with flexible framework models that allow for customization of processes to a particular project. Methodologies do have disadvantages and companies that see one and only one methodology, namely their own, may have a difficult time recognizing the pitfalls. But management consultants that have had the exposure to a variety of methodologies in various industries often have a much better understanding of the limitations.

6.7 LIFE-CYCLE PHASES

Determining the best number of life-cycle phases can be difficult when developing a project management methodology. As an example, let us consider IT. During the 1980s, with the explosion in software, many IT consulting companies came on the scene with the development of IT methodologies using SDLC phases. The consultants promise their client phenomenal results if the client purchases the package along with the accompanying training and consulting efforts. Then, after spending hundreds of thousands of dollars, the client reads the fine print that states that the methodology must be used as is, and no customization of the methodology will take place. In other words, the client must change their company to fit the methodology rather than vice versa. Most of the IT consultancies that adopted this approach no longer exist.

For an individual company, agreeing on the number of life-cycle phases may be difficult at first. But when an agreement is finally reached, all employees should live by the same phases. However, for today's IT consulting companies, the concept of one package fits all will not work. Whatever methodology they create must have flexibility in it so that client customization is possible. In doing so, it may be better to focus on processes rather than phases or possibly a framework approach that combines the best features of each.

Historically, we defined the first phase of a project as the initiation phase. This phase included bringing the project manager on board, handing him or her a budget and a schedule, and telling them to begin project execution. Today, there is a preinitiation phase which Russ Archibald and his colleagues refer to as the project incubation/feasibility phase.[8] In this phase, we look at the benefits of the project, the value expected at completion, whether sufficient and qualified resources are available, and the relative importance of the project compared to other projects that may be in the queue. It is possible that the project may never reach the initiation phase.

In the past, project management was expected to commence at the initiation phase because it was in this phase that the project manager was assigned. Today, project managers are expected to possess a much greater understanding of the business as a whole and companies have found it as beneficial to bring the project manager on board earlier than the initiation phase to assist in making business decisions rather than purely project decisions.

In the same context, we traditionally viewed the last life-cycle phase as project closure. This includes the implementation of contractual closure, administrative closure, and financial closure. After closure, the project manager would be reassigned to another project.

Today, we are including a post–project evaluation phase. Some companies refer to this as a customer satisfaction management phase. In this phase, selected members of the project team and sales/marketing personnel, as well as members from the governance committee, meet with the client to see what changes can be made to the methodology or processes used to execute the project and what can be done differently on future projects for this client to improve even further the working relationship between the client, contractor, and stakeholders.

6.8 DRIVERS FOR PM 2.0 CLIENT-CENTERED FLEXIBILITY

When you manage projects for internal clients, there may or may not be any need for flexibility in the application of the methodology. But for external clients, flexibility is mandatory. Clients want your methodology to fit their business model. If your methodology has different gates, different life-cycle phases, and other rigid requirements, the client may look for different providers for future work. The need for flexibility is based upon:

- Your understanding of the client's business
- The ability to plan for a continuously changing or moving target

---

[8] R. D. Archibald, I. Di Filippo, and D. Di Filippo have written an excellent paper on this topic, "The Six-Phase Comprehensive Project Life Cycle Model Including the Project Incubation/Feasibility Phase and the Post-Project Evaluation Phase," *PM World Journal*, December 2012.

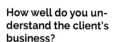

How well do you understand the client's business?

Copyright © Scott Maxwell/Fotolia

- The ability to align your life-cycle phases with those of the client
- The ability to revalidate the client's assumptions as well as yours
- The ability to perform trade-offs that satisfy both you and the client
- The ability to have a mutually agreed-upon definition of project success
- The ability to establish metrics that the client understands and can use for decision making

The closer you can get to the client's business model, the greater the customer satisfaction. But along the way there will be obstacles. The project managers using PM 2.0 principles are expected to make both business and project decisions. For this to happen, the project managers must understand not only their own business but also the client's business. Project managers will always be faced with some dilemmas. The client may not be willing to share with you any proprietary information concerning their strategy and you are still expected to make decisions that are aligned with their strategy. You may be restricted as to whom you can talk to in the client's company. You may have some people from the client's company interfacing with your project team and you may have to prove yourself to them before they trust you. Unfortunately, trust takes time and this is not a luxury for most project managers.

The first project with a new client is usually the most difficult project. If the client provides you with follow-on work and a trusting relationship develops, you may eventually be treated as a strategic partner as discussed in Chapter 2. In a best case scenario, the client may even treat you as an employee of their company and give you access to strategic information. But in any event, you must gather as much information as possible about the client's business model and future plans.

6.9 UNDERSTANDING MOVING TARGETS

When we manage projects, we assume that the SOW is well defined and will not change over the life of the project. In a PM 2.0 world, we must be able to live with moving targets. The longer the project, the greater the chance that the target might move because:

- The client's business base will change.
- The enterprise environmental factors have changed.
- Many of the assumptions have changed or are no longer valid.
- There are new constraints imposed upon the project.
- The priorities of the constraints have changed.

How do we handle continuously moving targets?

Copyright © Scott Maxwell/LuMaxArt/Shutterstock

Project managers must expect the worst and be willing to continuously replan the project if necessary. The hard part of replanning is to minimize the damage to the project while maximizing what can be salvaged. And, as expected, the client wants all of this accomplished within the original budget and schedule.

Today, we understand the necessity for the client to change the business case and the SOW and then request a moving target. More serious issues occur when the governance committee constantly changes the target, many times because of personal desires rather than based upon a sound need. This situation can be further compounded if the membership of the governance committee continuously changes.

6.10 NEED FOR CLIENT-SPECIFIC METRICS

In Chapter 2, we stated that the customer and contractor would come to some understanding on the definition of project success and then look in the contractor's metric library for supporting metrics. But what happens if the metrics that the client wants are not in the contractor's metric library?

The project manager must then make two critical assessments. First, how important is it to have these metrics as part of the methodology reporting system? Second, what is the additional cost, if any, to track, measure, and report these additional metrics? If the metrics are a necessity for alignment with the client's business strategy, then the metrics must be included in the methodology. The project manager needs to understand the new metrics, how they will be used, and especially how the client plans to use the metrics as part of informed decision making.

While all of this may sound good, the PMO must fully understand what is happening. The PMO must address the following:

How do we handle client-specific metrics?

Copyright © Scott Maxwell/Fotolia

- The PMO performs routine health checks on projects as well as audits. The PMO must understand that client-specific metrics are being used.
- The PMO usually manages the metrics library. The PMO must understand that these metrics that are client-specific may be considered by the client as company proprietary knowledge and should not be added into the metrics library at the completion of the project.
- The use of the additional metrics may necessitate that a confidentiality agreement be signed when using some company-specific metrics.

The use of client-specific metrics is a stepping-stone to customer satisfaction and repeat business. While there may be headaches along the way, this is a necessity. A lot of these issues can be resolved early on in the project if we have an engagement project manager as discussed in Chapter 2 working alongside the project manager.

6.11 BUSINESS CASE DEVELOPMENT

One of the characteristics of PM 2.0 is that project managers are brought on board perhaps as early as the portfolio selection stage. As such, project managers may be actively involved in the preparation of the business case rather than taking it for granted that someone else prepared it and that it is correct and realistic. For external clients, project managers may be expected to understand the client's business model well enough to assist the client in the preparation of the business case.

A business case is a document that provides the reasoning why a project should be initiated. Historically, business cases were small documents or presentations and the decision to initiate the project was based upon the rank of the person making the request. Today, business cases are well-structured written documents that support a specific business need. Each business case should describe the boundaries to the project in sufficient detail such that the decision makers can determine that the expected business value and benefits exceed the cost of performing the project.

The business case must contain both quantifiable and unquantifiable information that justifies the investment in the project. Typical information that can be part of a business case includes:

- **Business Need:** This identifies the gap that currently exists and the need for the investment.
- **Opportunity Options:** This identifies how the project is linked to strategic business objectives.
- **Benefit Realization Plan:** This identifies the value/benefits (rather than products or deliverables) that can be obtained whether they are cost savings, additional profits, or opportunities.
- **Assumptions Made:** This identifies all of the assumptions that are made to justify the project.
- **Recommendation for Evaluation:** This identifies what techniques should be used for evaluation, such as a benefit-to-cost ratio, cash flow considerations, strategic options, opportunity costs, return on investment, net present value, and risks.
- **Project Metrics:** This identifies the financial and nonfinancial metrics that will be used to track the performance of the project.
- **Exit Strategies:** This identifies the criteria to be used to cancel the project if necessary.
- **Project Risks:** This helps the decision makers evaluate the project by listing briefly the business, legal, technical, etc., risks of the project.
- **Project Complexity:** This identifies how complex the project might be, perhaps even from a risk perspective, if the organization can manage the complexity, and if it can be done with existing technology.
- **Resources Needed:** This identifies the human and nonhuman resources needed.
- **Timing:** This identifies the major milestones for the project.
- **Legal Requirements:** This identifies any legal requirements that must be followed.

The above information is used not only to approve the project but also to be able to prioritize it with all of the other projects in the queue.

The person preparing the business case may not have all of the above information. It may be necessary to have either the project manager or the PMO involved in preparing the business case prior to submission.

6.12 VALIDATING ASSUMPTIONS

Planning begins with an understanding of the assumptions. Project planning is based upon the expectation that future results can be extrapolated from past experiences. If experience is lacking or extrapolation will generate misleading information, then assumptions must be made to predict future outcomes.

Quite often, the assumptions are made by marketing and sales personnel and then approved by senior management as part of the project selection and approval process. The expectations for the final results are based upon the assumptions made.

Why is it that, more often than not, the final results of a project do not satisfy senior management's expectations? At the beginning of a project, it is impossible to ensure that the benefits expected by senior management will be realized at project completion. While project length is a critical factor, the real culprit is changing assumptions.

Assumptions can be made for items that are or are not under the direct control of the project team but can influence the outcome of the project. Assumptions made by the project manager are often surrounding the enterprise environment factors and the organizational process assets.

Enterprise Environmental Factors: These are assumptions about external environmental conditions that can affect the success of the project, such as interest rates, market conditions, changing customer demands and requirements, customer involvement, changes in technology, political climate, and even government policies.

Organizational Process Assets: These are assumptions about present and future company assets that can impact the success of the project, such as the capability of your project management methodology, the project management information system, forms, templates, guidelines, checklists, the ability to capture and use lessons learned data and best practices, resource availability, and skill level.

At the onset of the project, all assumptions must be challenged to verify their validity. As the project progresses, the assumptions must be tracked and revalidated. As described earlier in this chapter, there are numerous reasons, both business and technical, as to why assumptions can change. If the assumptions change or are no longer valid, then perhaps the project should be redirected or even canceled. Unfortunately, many project managers do not track the validity of the assumptions and end up completing a project within time and cost, but the project does not add business value to the company or the client. PM 2.0 mandates that all assumptions be tracked for possible changes.

It may be impossible to make assumptions with any degree of accuracy. However, this should not prevent risk management from taking place on the assumptions and contingency plans established if the assumptions are proven to be false.

Types of Assumptions

There are several types of assumptions. The two most common categories are explicit and implicit assumptions and critical and noncritical assumptions. Critical and noncritical assumptions are also referred to as primary and secondary assumptions. These two categories of assumptions are not mutually exclusive.

Explicit assumptions may be quantified and are expressed without any ambiguity. Implicit assumptions may be hidden and may go undetected. Explicit assumptions often contain hidden implicit assumptions. As an example, we could make an explicit assumption that five people will be needed full time to complete the project. Hidden is the implicit assumption that the people assigned will be available full time and possess the necessary skills. Serious consequences can occur if the implicit assumptions are proven to be false.

Critical assumptions are those assumptions that can cause significant damage to a project if even small changes take place. Critical assumptions must be tracked closely whereas noncritical assumptions may not be tracked and may not require any action as long as they do not become critical assumptions. Project managers must develop a plan for how they will measure, track, and report the critical assumptions. Measurement implies that the assumptions can be quantified. Since assumptions predict future outcomes, testing and measurement may not be able to be made until well into the future or unless some risk triggers appear. Sensitivity analysis may be required to determine the risk triggers.

In agile fixed-price contracts, the project manager and the customer work together to identify the assumptions. An agreement must be reached on the critical assumptions, especially with regard to business value, risks, and costs. An understanding must also be reached on what changes to the critical assumptions may trigger the need for scope changes. This requires that the project manager and the customer remain in close collaboration throughout the life of the project.

There are some assumptions that project managers may never see or even know about. These are referred to as strategic assumptions and are held by decision makers when approving a project or selecting a portfolio of projects. These types of assumptions may contain company proprietary information that executives do not want the project team to know about.

Documenting Assumptions

Assumptions must be documented at project initiation using the project charter as a possible means. Throughout the project, the project manager must revalidate and challenge the assumptions. Changing assumptions may mandate that the project be terminated or redirected toward a different set of objectives.

A project management plan is based upon the assumptions described in the project charter. But there may be additional assumptions made by the team that are inputs to the project management plan. One of the primary reasons why companies use a project charter is that project managers were most often brought on board well after the project selection process and approval process were completed. As a result, project managers needed to know what assumptions were considered.

Enterprise Environmental Factors: These are assumptions about the external environmental conditions that can affect the success of the project, such as interest rates, market conditions, changing customer demands and requirements, changes in technology and even government policies.

Organizational Process Assets: These are assumptions about present or future company assets that can impact the success of the project such as the capability of your enterprise project management methodology, the project management information system, forms, templates, guidelines, checklists, and ability to capture and use lessons learned data and best practices.

Documenting assumptions is necessary in order to track the changes. Examples of assumptions that are likely to change over the duration of a project, especially on a long-term project, might be that:

- The cost of borrowing money and financing the project will remain fixed.
- The procurement costs will not increase.

TABLE 6-1 Assumption Validation Checklist		
Checklist for Validating Assumptions	**Yes**	**No**
Assumption is outside of the control of the project		
Assumption is outside of the control of the stakeholder(s)		
The assumption can be validated as correct		
Changes in the assumption can be controlled		
The assumed condition is not fatal		
The probability of the assumption holding true is clear		
The consequences of this assumption pose a serious risk to the project		
Unfavorable changes in the assumption can be fatal to the project		

- The breakthrough in technology will take place as scheduled.
- The resources with the necessary skills will be available when needed.
- The marketplace will readily accept the product.
- Our competitors will not catch up to us.
- The risks are low and can be easily mitigated.
- The political environment in the host country will not change.

The problem with having faulty assumptions is that they can lead to faulty conclusions, bad results, and unhappy customers. The best defense against poor assumptions is good preparation at project initiation, including the development of risk mitigation strategies. One possible way to do this is with a validation checklist as shown in Table 6-1.

6.13 DESIGN FREEZES

A design freeze is a point in a project where no further changes to the design of the product can be made without incurring a financial risk and/or a schedule slippage. The decision points for the freeze usually occur at the end of specific life-cycle phases. There are several types of freezes, and they can occur in just about any type of project. However, they are most common in new product development (NPD) projects.

In NPD projects, we normally have both a specification freeze and a design freeze. The specification dictates the set of requirements upon which the final design must be made. Following a specification freeze, we have a design freeze whereby the final design is handed to manufacturing. The design freeze may be necessary for timely procurement of long lead items such as parts and tooling that are necessary for the final product to be manufactured. The timing of the design freeze is often dictated by the lead times and may be beyond the control of the company. Failing to meet design freeze points has a significantly greater impact on manufacturing than engineering design.

Design freezes have the additional benefit of controlling downstream scope changes. However, even though design changes can be costly, they are often necessary for

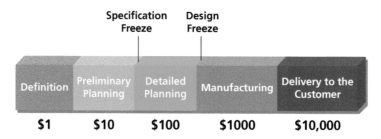

FIGURE 6-1   Cost of corrections

safety reasons, to protect the firm against possible product liability lawsuits, and to satisfy a customer's specific needs.

Changes to a product after the product's design is handed over to manufacturing can be costly. As a rule, we generally state that the cost of a change in any life-cycle phase after the design freeze is 10 times the cost of performing the change in the previous life-cycle phase. This is shown in Figure 6-1. If a mistake is made prior to the design freeze point, the correction could have been made for $100. But if the mistake is not detected until the manufacturing stage, then the correction cost could be $1000. The same mistake, if detected after the customer receives the product, could cost $10,000 to correct a $100 planning mistake or a $1000 manufacturing mistake. While the rule of 10 may seem a little exaggerated, it does show the trend in correcting costs downstream.

We seem to have a good idea about design freeze impacts on new product development. But for projects where an agile methodology is used and the requirements may be continuously evolving, the decision on design freeze points may be difficult to implement. This will be a challenge that we will need to overcome with PM 2.0. It will be necessary for the customer and contractor to work together closely and identify the accompanying risks.

6.14 CUSTOMER APPROVALS

Project managers often neglect to include milestones or time durations for customer approvals in the project's schedule baseline with the mistaken belief that the approval process will happen quickly. The approval process in the project manager's parent company may be known with some reasonable degree of certainly, but the same cannot be said for the client's approval process. For PM 2.0 projects where there can be several stakeholders and virtual project management teams, the approval process can be quite long. Good project management metrics and dashboards may be able to alleviate the problems.

Factors that can impact the speed by which the approvals take place can include:

- How many people are involved in the approval process
- Whether any of the people are new to the project
- When all of the necessary participants can find a mutually agreed-upon time to meet
- The amount of time the people need to review the data, understand the data, and determine the impact of their decision

- Their knowledge of project management
- A review of the previous project decisions that were made by them or others
- How well they understand the impact of a delayed decision
- Whether they need additional information to make a decision
- Whether the decision can be made verbally or if it must be written in a report format

Simply adding a milestone to a schedule that says "customer approval" does not solve the problem. Project managers must find out how long their customers need to make a decision, and it may be better to indicate customer approval as an activity rather than as a milestone. For projects based upon an agile methodology or projects that require a rapid development process, we must make sure early on in the project that all decision makers understand the impact of late decisions.

6.15 AGILE PROJECT MANAGEMENT METHODOLOGY

When project management practices began to be used, companies were unsure as to whether they would work. Executives wanted a methodology created whereby there would be standardization and consistency in the way that projects would be planned, scheduled, monitored, and controlled. As a result, methodologies were based upon rather rigid policies and procedures that were deemed necessary to meet the inflexible budget and schedule constraints. As long as the scope was reasonably well defined, it was not often difficult to meet the fixed-cost and schedule constraints as long as the budget and schedule were realistic. These traditional methodologies had structured life-cycle phases for customer and stakeholder involvement.

As project management began to mature, executives and customers were expressing more trust in the project managers. Project managers were then given more freedom in how to apply the project management methodology. Policies and procedures were replaced with forms, guidelines, templates, and checklists. Project managers now had the freedom to decide which processes should be used on the project.

As the use of project management began to expand throughout the enterprise, project management practices were being applied to nontraditional projects. Some of the characteristics of these nontraditional projects are:

- The outcome of the project could change.
- With a possibly changing outcome, numerous scope changes would be expected.
- Customer and stakeholder involvement would need to occur more frequently rather than at the end of each life-cycle phase.
- The budget and schedule were no longer fixed.
- Success was now being measured by the value that would be created rather than just in terms of time and cost.
- The project team must possess innovation skills.

Agile project management techniques were created to satisfy these nontraditional characteristics which are quite common on IT projects. Not all projects can use the agile approach and not all customers are willing to accept a more flexible and value-driven approach at the expense of having a fixed cost and schedule.

Waterfall (Traditional and Modified)   **Agile**

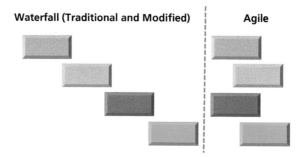

FIGURE 6-2 Waterfall versus agile charts

On agile projects, change can occur quickly, thus necessitating continuous customer involvement. Continuous customer involvement can minimize the documentation efforts. Project management flexibility is essential. Potential scope changes and continuous replanning efforts help mitigate the project's risks and prevent disasters. The agile methodology allows the project team to react quickly to a changing business environment and to respond to changing or evolving needs.

Traditional project management methodologies are normally shown as a waterfall chart where all of the life-cycle phases are performed sequentially. This is shown in Figure 6-2. Based upon the amount of risk the project manager is willing to accept, some of the life-cycle phases can overlap, thus giving us a modified waterfall chart. Waterfall charts imply structure, control, finite life-cycle phases, and well-defined decision points. With agile, work can be conducted in parallel, thus requiring more replanning and increased customer involvement.

In Chapter 1, we stated that more contracts associated with PM 2.0 would be cost-reimbursable contracts because of the complexity, risks, and uncertainties associated with a PM 2.0 environment. However, in agile development, especially for IT, the contracts are fixed price because of the cooperation needed to balance the interests of the contractors and the customers. Agile fixed-price contracts are cooperative contracts where the schedule and cost may be fixed and the scope is continuously evolving. In the traditional waterfall approach, scope is fixed or well defined up front and the cost and schedule may be allowed to be modified as the project continues. According to Andreas Opelt[9]:

> The main characteristic of a shift to the agile paradigm is that the scope of an IT project is in contrast to the classical waterfall model, no longer fixed in detail from the start. Instead, costs and time are defined on the basis of principles agreed over the course of the project, and the scope of services is developed and implemented step by step in short iteration cycles. This means that the assumption of a detailed prediction of the scope is omitted. For this model to be reflected in the contract, an agile fixed-price contract does not define an exact scope but, rather, supports an empirical process.
>
> An agile fixed-price contract defines a contractual framework within which time and cost are agreed upon as well as a structured approach to steer the scope within boundaries and by processes in a defined and controlled manner. Thus, the contractual model of an agile

[9] A. Opelt, B. Gloger, W. Pfarl, and R. Mittermayr, *Agile Contracts*, Wiley, Hoboken, NJ, 2013, pp. 47–48.

fixed-price contract reacts to two uncertainties. On the one hand, you never know exactly what details will be needed at the start of a project. On the other hand, you do not always need everything that had originally been considered to be important. This may only become apparent during the project's progress or perhaps only after project completion. An agile fixed-price contract therefore includes scope control, which makes a decision possible as to whether a specific feature should be more or less complex or whether it needs to be produced at all during the actual development process.

This does not mean, however, that customers have no idea what they are getting for their money at the start of a project. It means that customers know right from the beginning what they will have to spend to meet the business requirements, which have been defined at a moderate level of detail at the start of a project.

6.16 IMPLEMENTING METHODOLOGY

The physical existence of a methodology does not convert itself into a world-class methodology. Methodologies are nothing more than pieces of paper. What converts a standard methodology into a world-class methodology is the culture of the organization and the way the methodology is implemented and used.

The existence of a world-class methodology does not by itself constitute excellence in project management. The corporatewide acceptance and use of it do lead to excellence. It is through excellence in execution that an average methodology becomes a world-class methodology.

One company developed an outstanding methodology for project management. About one-third of the company used the methodology and recognized its true long-term benefits. The other two-thirds of the company would not support the methodology. The president eventually restructured the organization and mandated the use of the methodology.

The importance of execution cannot be underestimated. One characteristic of companies with world-class project management methodologies is that they have world-class managers throughout their organization.

Rapid development of a world-class methodology mandates an executive champion, not merely an executive sponsor. Executive sponsors are predominantly on an as-needed basis. Executive champions, on the other hand, are hands-on executives who drive the development and implementation of the methodology from the top down. Most companies recognize the need for the executive champion. However, many companies fail to recognize that the executive champion position is a life-long experience. One Detroit company reassigned its executive champion after a few successes were realized using the methodology. As a result, no one was promoting continuous improvement to the methodology.

Good project management methodologies allow you to manage your customers and their expectations. If customers believe in your methodology and it is a good fit with the customer's business model, then they usually understand it when you tell them that no further scope changes are possible once you enter a specific life-cycle phase. One automotive subcontractor carried the concept of trust to its extreme. The contractor invited the customers to attend the contractor's end-of-phase review meetings. This fostered extreme trust between the customer and the contractor. However, the customer was asked to leave during the last 15 min of the end-of-phase review meetings when project finances were being discussed.

Project management methodologies are an "organic" process, which implies that they are subject to changes and improvements. Typical areas of methodology improvement might include:

- Improved interfacing with suppliers
- Improved interfacing with customers
- Better explanation of subprocesses
- Clearer definition of milestones
- Clearer role delineation of senior management
- Recognition of need for additional templates
- Recognition of need for additional metrics
- Template development for steering committee involvement
- Enhancement of the project management guidebook
- Ways to educate customers on how the methodology works
- Ways of shortening baseline review meetings

6.17 IMPLEMENTATION BLUNDERS

Even though companies recognize the driving forces which indicate a need for project management improvement, the actual decision to make an investment to do it may not happen until some crisis occurs or a significant amount of red ink appears on the company's balance sheet. Recognizing a need is a lot easier than doing something about it because doing it requires time and money. Too often, executives procrastinate giving the go-ahead in hopes that a miracle will occur and project management improvements will not be necessary. And while they procrastinate, the situation often deteriorates further.

Delayed investment in project management capabilities such as in PM 2.0 is just one of many blunders. Another common blunder, which can occur in even the best companies, is the failure to treat project management as a profession. In some companies, project management is a part-time activity to be accomplished in addition to one's primary role. The career path opportunities come from the primary role, not through project management. In other companies, project management may be regarded merely as a specialized skill in the use of scheduling tools.

6.18 OVERCOMING DEVELOPMENT AND IMPLEMENTATION BARRIERS

Making the decision that the company needs a project management methodology is a lot easier than actually doing it. There are several barriers and problems that surface well after the design and implementation team begins their quest. Typical problem areas include:

- Should we develop our own methodology or benchmark best practices from other companies and try to use their methodology in our company?

- Can we get the entire organization to agree upon a singular methodology for all types of projects or must we have multiple methodologies?
- If we develop multiple methodologies, how easy or difficult will it be for continuous improvement efforts to take place?
- How should we handle a situation where only part of the company sees a benefit in using this methodology and the rest of the company wants to do its own thing?
- How do we convince the employees that project management is a strategic competency and the project management methodology is a process to support this strategic competency?
- For multinational companies, how do we get all worldwide organizations to use the same methodology? Must it be intranet based?

These are typical questions that plague companies during the methodology development process. These challenges can be overcome, and with great success, as illustrated by the companies identified in the following sections.

6.19 USING CRISIS DASHBOARDS WITH METHODOLOGIES

While methodologies serve a viable purpose, even flexible methodologies may not work well when projects become distressed and rapid action is required to save a failing project. Unless the methodologies somehow identify the crisis situations, delays in solving problems can happen.

Over the past several years, dashboards have become commonplace for presenting project status information to the project team, clients, and stakeholders. The purpose of a dashboard is to convert raw data into meaningful information that can easily be understood and used for informed decision making. The dashboard provides the viewer with "situational awareness" of what the information means now and what it might mean in the future if the existing trends continue. Dashboards function as communication tools that allow us to go to paperless project management and fewer meetings as well as eliminate waste.

Projects in today's environment are significantly more complex than projects managed in the past. With today's projects, governance is performed by a governance committee rather than just a project sponsor. Each stakeholder or member of the governance committee may very well require different metrics and KPIs. If each stakeholder wishes to view 20–30 metrics, the costs of metric measurement and reporting can be significant and can defeat the purpose of going to paperless project management.

The solution to effective communications with stakeholders and governance groups is to show them that they can most likely get all of the critical data they need for informed decision making with 6–10 metrics or KPIs, which can be displayed on one computer screen. This is not always the case, however, and drilling down to other screens may be necessary. But, in general, one computer screen shot should be sufficient.

If an out-of-tolerance condition or crisis situation exists with any of the metrics or KPIs on the dashboard screen, then the situation should be readily apparent to the viewer with the crisis situation being highlighted. But what if the crisis occurs due to metrics that do not appear on the screen? In this case, the viewer will be immediately directed to a crisis dashboard which shows all of the metrics that are out of tolerance.

The metrics will remain on the crisis dashboard until such time as the crises or out-of-tolerance conditions are corrected. Each stakeholder will now see the regular screen shot and then be instructed to look at the crisis screen shot.

Understanding Targets

Metrics and KPIs are measurements in relation to a target. The metric or KPI tells us how far we are above or below the predefined target. Meeting the target implies that we are giving the stakeholders the value they expect from the project.

Typical targets for a metric or KPI might be:

- Simple quantitative targets
- Time-based targets that are measured monthly or during a certain time interval
- At completion targets that are measured at work package or project completion
- Stretch targets, such as becoming best in class or performance greater than specification requirements
- Visionary targets that may be well into the future, such as more repeat business from this client

Simple quantitative targets are the most common. Examples of these are shown in Table 6-2. Targets must be realistic and not necessarily challenging. Otherwise, workers might try to circumvent the targets. Targets may require trial-and-error solutions.

Hitting a target exactly requires some degree of luck. Missing a target may be acceptable if the variance from the target, whether favorable or unfavorable, falls within acceptable limits. The limits are referred to as the tolerances, thresholds, or the integrity of the target. Therefore, when establishing a target for each metric, it is important to also establish the limits. The established limits must be acceptable to the project team, the client, and the stakeholders. Typical limits might be the target ±5% or the target ±10%.

TABLE 6-2 Examples of Simple Targets

Type of Target	Example
A single value	Completion of 20 tests
An upper limit	≤$200,000
A lower limit	≥$100,000
A range of values	$400,000 ± 10%
A percentage of a specific quantity that may be fixed	Scrap factors are not more than 5% of material costs
A percentage of a specific activity that may change	Planning dollars are not more than 35% of total labor dollars
Accomplishment milestones and deliverables	Must produce and ship at least 10 deliverables per month

The magnitude of the limits is often based upon the accuracy of the measurement techniques to be used. Poor measurement techniques may justify larger limits. However, some companies maintain EPM methodologies that define the tolerances for each metric. This occurs mainly in organizations reasonably mature in project management and with some experience in metrics management. It is also possible, though uncommon, for the business case of the project to identify the critical metrics, the targets, and the tolerances.

Defining a Crisis

A crisis can be defined as any event, whether or not expected, that can lead to an unstable or dangerous situation affecting the outcome of the project. Crises imply negative consequences that can harm the organization, its stakeholders, and the general public. Crises can cause changes to the firm's business strategy, how it interfaces with the enterprise environmental factors, the firm's social consciousness, and the way it maintains customer satisfaction. A crisis does not necessarily mean that the project will fail, nor does it mean that the project should be terminated. The crisis could simply be that the project's outcome will not occur as expected.

Some crises may appear gradually and can be preceded by early warning signs. These are referred to as "smoldering" crises. The intent of metrics and dashboards is to identify trends that could indicate that a crisis is coming and to provide the project manager with sufficient time to develop contingency plans and take corrective action. The earlier you know about the impending crisis, the more options you may have available as a remedy.

Another type of crisis is that which occurs abruptly with little or no warning. These are referred to as "sudden" crises. Examples which could impact projects might be elections or political uncertainty in the host country, natural disasters, or the resignation of an employee with critical skills. Metrics and dashboards cannot be created for every possible crisis that could exist on a project. Sudden crises cannot be prevented.

Not all out-of-tolerance conditions are a crisis. For example, being significantly behind schedule on a software project may be seen as a problem but not necessarily a crisis. However, if you are behind schedule on the construction of a manufacturing plant and plant workers have already been hired to begin work on a certain date or the delay in the plant will activate penalty clauses for late delivery of manufactured items for a client, then this could constitute a crisis. Sometimes, exceeding a target favorably may also trigger a crisis. A manufacturing company had a requirement to deliver 10 and only 10 units to a client each month. The company manufactured 15 units each month but could only ship 10 per month to the client. Unfortunately, the company did not have storage facilities for the extra units produced, and a crisis occurred.

When establishing targets, we must identify which of the out-of-tolerance conditions will trigger a crisis and which represent just a problem. In Figure 6-3, exceeding the tolerances of the target ±10% unfavorably is identified as a potential problem. In Figure 6-4, we have identified thresholds for both a problem and a crisis.

How do we determine whether the out-of-tolerance condition is just a problem or a crisis? The answer is in the potential damage that can occur. If any of the following can occur, then the situation would most likely be treated as a crisis:

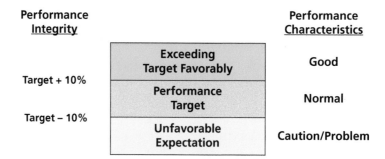

FIGURE 6-3   Threshold limits for a problem

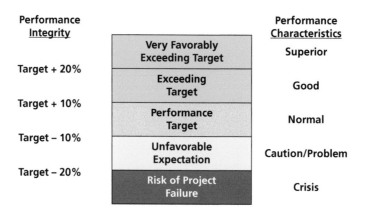

FIGURE 6-4   Threshold limits for a problem and a crisis

There is a significant threat to the outcome of the project.

- There is a significant threat to the organization as a whole, its stakeholders, and possibly the general public.
- There is a significant threat to the firm's business model and strategy.
- There is a significant threat to worker health and safety.
- There is a possibility for loss of life.
- Redesigning existing systems is now necessary.
- Organizational change will be necessary.
- The firm's image or reputation will be damaged.
- Degradation in customer satisfaction could result in a present and future loss of significant revenue.

It is important to understand the difference between risk management and crisis management. According to Wikipedia, the free encyclopedia:

In contrast to risk management, which involves assessing potential threats and finding the best ways to avoid those threats, crisis management involves dealing with threats before, during, and after they have occurred. That is, crisis management is proactive, not merely reactive. It is a discipline within the broader context of management consisting of skills and techniques required to identify, assess, understand, and cope with a serious situation, especially from the moment it first occurs to the point that recovery procedures start.

Crises often require that immediate decisions be made. Effective decision making requires information. If one metric appears to be in a crisis mode and shows up on the crisis dashboard, the viewers may find it necessary to look at several other metrics which may not be in a crisis mode and may not appear on the crisis dashboard but are possible causes of the crisis. Looking at metrics on dashboards is a lot easier than reading reports.

The difference between a problem and a crisis is like beauty; it is in the eyes of the beholder. What one stakeholder sees as a problem another stakeholder may see as a crisis. Table 6-3 shows how difficult it is to make the differentiation.

TABLE 6-3 Differentiating between a Problem and a Crisis

Metric/KPI	Problem	Crisis
Time	The project will be late but still acceptable to the client.	The project will be late and the client is considering cancellation.
Cost	Costs are being overrun but the client can provide additional funding.	Costs are being overrun and no additional funding is available. Cancellation is highly probable.
Quality	The customer is unhappy with the quality but can live with it.	Quality of the deliverables is unacceptable, personal injury is possible, the client may cancel the contract, and no further work may come from this client.
Resources	Either the project is understaffed or the resources assigned have marginal skills to do the job. A schedule delay is probable.	The quality or lack of resources will cause a serious delay in the schedule, and the quality of workmanship may be unacceptable such that the project may be canceled.
Scope	There are numerous scope changes which cause changes to the baselines. Delays and cost overruns are happening but are acceptable to the client for now.	The number of scope changes has led the client to believe that the planning is not correct and more scope changes will occur. The benefits of the project no longer outweigh the cost, and project termination is likely.
Action items	The client is unhappy with the amount of time taken to close out action items, but the impact on the project is small.	The client is unhappy with the amount of time taken to close out action items, and the impact on the project is significant. Governance decisions are being delayed, and the impact on the project may be severe.
Risks	Significant risk levels exist, but the team may be able to mitigate some of the risks.	The potential damage that can occur because of the severity of the risks is unacceptable to the client.
Assumptions and constraints	New assumptions and constraints have appeared and may adversely affect the project.	New assumptions and constraints have appeared. The value of the project may no longer be there.
Enterprise environmental factors	The enterprise environmental factors have changed and may adversely affect the project.	The new enterprise environmental factors will greatly reduce the value and expected benefits of the project.

Crisis Dashboard Images

There are three important facts that must be remembered about dashboard images. First, the way that the image appears on a crisis dashboard can be different from the way it appears on a traditional dashboard in order to make it clear that a crisis may exist. Second, there is a fine line between what is or is not a crisis. Third, the viewer of the crisis dashboard is the person who ultimately makes the decision as to whether the situation is a crisis or just a problem, even though we may have exceeded the threshold limits for a crisis.

Let us assume that you price out a project, and the functional managers inform you that the requirements of the project mandate that workers in the grade 7 and grade 8 categories be assigned to perform the work. A grade 8 worker is higher skilled than a grade 7 worker. From Figure 6-5, we can see that there are several people assigned who are grade 6 workers, and they may not be qualified to perform the necessary work.

The viewer of the dashboard image in Figure 6-5 might not realize that this may be a problem or a crisis. But if the image is reconfigured to what you see in Figure 6-6, it is easier to see that a crisis may exist. The bold horizontal line indicates that no more than 20% of the work force should be below the minimum acceptable pay grade without considering it a possible crisis. The numbers in the columns indicate the number of workers assigned who are below the minimum acceptable level. Even though a minimum acceptable level has been established, it is possible that some stakeholders may not consider this a crisis. They may not even consider it as a problem if the work is being accomplished correctly.

Sometimes, because of limited space on the crisis dashboard, we find it necessary to combine images. As an example, on routine dashboards we often create separate images for deliverables produced on time or late and for deliverables accepted or rejected by the customer. When they are combined for a crisis dashboard, they may appear as shown in Figure 6-7. Assume that the company considers one deliverable a month to be rejected and returned for rework as acceptable. This is the situation for January and February.

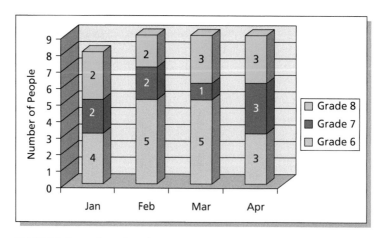

FIGURE 6-5   Assigned workers by pay grade

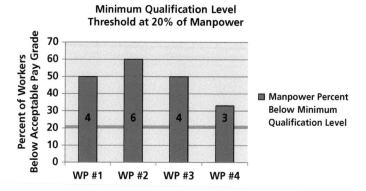

FIGURE 6-6 Manpower levels with thresholds

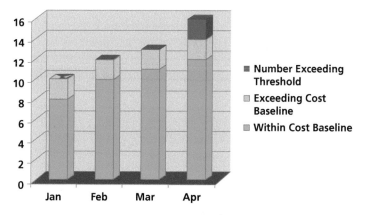

FIGURE 6-7 Accepted or rejected completed metrics

But in April two deliverables were rejected by the client, and this in turn triggered a possible crisis. Once again, the client may consider this just a problem rather than a crisis.

Stakeholders are interested more in where we will end up rather than where we are today. As such, the traditional dashboards that stakeholders look at should have metrics and KPIs that can be extrapolated into the future. In other words, the cost variance (CV) and the schedule variance (SV) are not as important to the stakeholders as are the cost performance index (CPI) and the schedule performance index (SPI). Establishing thresholds for a crisis is easier with CPI and SPI than with CV and SV. Figure 6-8 is an example where the crisis thresholds were set at 0.80 for both SPI and CPI.

Sometimes the trend is more important than the absolute value of the indicator at a specific point in time. As an example, consider how important risk management has become in recent years. There is a tendency today to assign a risk value perhaps on all of the work packages in a project. Risks are normally the greatest at the beginning of a project, because risks and knowledge are inversely related. The more knowledge you have, the less the risk. Therefore, as the project develops and more knowledge is gained, we would expect many of the risks to be reduced.

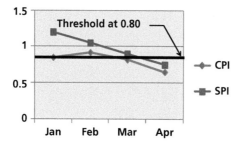

FIGURE 6-8 **Cumulative month-end CPI and SPI data**

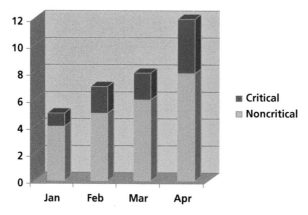

FIGURE 6-9 **Work packages open with and without critical risk designation**

But what happens if the risks actually increase as shown in Figure 6-9? In this case, it is the trend that is important. The trend in Figure 6-9 shows that the number of work packages with a critical risk designation is increasing. This may not be a crisis but may simply indicate that this should be monitored. It is possible that this situation was expected.

Figure 6-9 could have been redrawn similar to Figure 6-4 and we could state that, if the number of work packages with a critical risk designation in any month exceeds 20%, then we have a crisis. But this can be confusing, because we could have just 3 work packages and only 1 is labeled as critical. This would not be a crisis. We could also have 100 work packages and 15 of them are considered critical. In this case, we would not reach the crisis trigger point of 20%, yet having 15 critical work packages in one month may actually be a crisis. In other words, some situations may not lend themselves to establishing a quantitative threshold value for a crisis trigger. The trends may be of more value than the actual numbers.

While critical thresholds for a risk crisis may not be practical at the work package level, the same does not hold true for total project risk. As an example, consider a company that identifies three primary project risks as shown in Figure 6-10: a delivery risk,

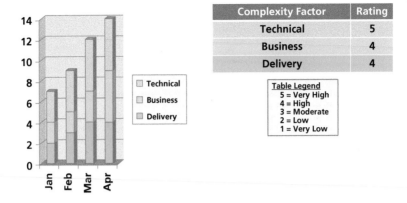

Complexity Factor	Rating
Technical	5
Business	4
Delivery	4

Table Legend
5 = Very High
4 = High
3 = Moderate
2 = Low
1 = Very Low

FIGURE 6-10   Project complexity (risk) factor

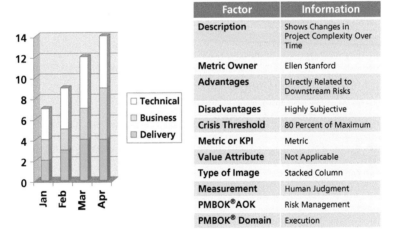

Factor	Information
Description	Shows Changes in Project Complexity Over Time
Metric Owner	Ellen Stanford
Advantages	Directly Related to Downstream Risks
Disadvantages	Highly Subjective
Crisis Threshold	80 Percent of Maximum
Metric or KPI	Metric
Value Attribute	Not Applicable
Type of Image	Stacked Column
Measurement	Human Judgment
PMBOK®AOK	Risk Management
PMBOK® Domain	Execution

FIGURE 6-11   Metric library representation for the project complexity (risk) factor shown in Figure 6-10

a business risk, and a technical risk. If each of these three risks were considered to be very high risks, then from the legend each would receive a value of 5, for a total score of 15. The company may then consider a crisis threshold for total project risk when the total score exceeds 80% of the maximum score possible. In other words, if the total score exceeds 80% × 15 maximum, or 12, the crisis threshold has been penetrated. In Figure 6-10 we see that the boundary of the crisis threshold was reached in March and exceeded in April.

The thresholds for a possible crisis are now being included in the metrics library maintained by the PMOs. An example of this can be seen in Figure 6-11 for the risk factor discussed above.

Conclusions

We can now draw the following conclusions about crisis dashboards:

- The definition of what is or is not a crisis is not always clear to the viewers.
- Not all problems are crises.
- Sometimes, unfavorable trends are treated as crises and appear on the crisis dashboards.
- The crisis dashboard may contain a mixture of crisis metrics and metrics that are treated as just problems.
- The metrics that appear on a traditional dashboard reporting system may have to be redrawn when placed on a crisis dashboard to make sure that the metrics are easily understood.
- Crisis metrics generally imply that either the situation must be monitored closely or some decisions must be made.

As we become more knowledgeable in crisis dashboards, it may eventually become more of a science than an art. But for now, since we are just in the infancy stages of understanding crisis dashboards, we must simply do the best we can.

6.20 SHUTTING DOWN THE PROJECT

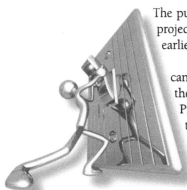

How difficult is it to shut down a project?

The purpose of having a crisis dashboard is to identify early warning signs that a project is in trouble in hope that recovery plans can be developed. But as stated in earlier chapters, not all projects can or should be saved.

Methodologies have decision points and off ramps whereby a project can be canceled. Unfortunately, the hardest decision for an executive to make is to pull the plug on a project and possibly to admit defeat or a poor initial decision. Project managers, sponsors, and stakeholders do everything possible to make their project successful. But what if the project champions, as well as the project team, have blind faith in the success of the project? What happens if the strongly held convictions and the collective belief disregard the early warning signs of imminent danger? What happens if the collective belief drowns out dissent?

In such cases, an exit champion must be assigned. The exit champion is usually someone from outside the project and at the upper levels of management who can perform a health check and make an impartial decision and presentation to the executive staff as to whether the project should continue on. The exit champion sometimes needs to have some direct involvement in the project in order to have credibility, but direct involvement is not always a necessity. Exit champions must be willing to put their reputation on the line and possibly face the likelihood of being cast out from the project team. According to Isabelle Royer[10]:

> Sometimes it takes an individual, rather than growing evidence, to shake the collective belief of a project team. If the problem with unbridled enthusiasm starts as an unintended

---

[10] I. Royer, "Why Bad Projects Are So Hard to Kill," *Harvard Business Review*, February 2003, p.11. Copyright © 2003 by the Harvard Business School Publishing Corporation. All rights reserved.

consequence of the legitimate work of a project champion, then what may be needed is a countervailing force—an exit champion. These people are more than devil's advocates. Instead of simply raising questions about a project, they seek objective evidence showing that problems in fact exist. This allows them to challenge—or, given the ambiguity of existing data, conceivably even to confirm—the viability of a project. They then take action based on the data.

The larger the project and the greater the financial risk to the firm, the higher up the exit champion should reside. If the project champion just happens to be the CEO, then someone on the board of directors or even the entire board of directors should assume the role of the exit champion. Unfortunately, there are situations where the collective belief permeates the entire board of directors. In this case, the collective belief can force the board of directors to shirk their responsibility for oversight.

Large projects incur large cost overruns and schedule slippages. Making the decision to cancel such a project, once it has started, is very difficult, according to David Davis[11]:

> The difficulty of abandoning a project after several million dollars have been committed to it tends to prevent objective review and recosting. For this reason, ideally an independent management team—one not involved in the projects development—should do the recosting and, if possible, the entire review. . . . If the numbers do not holdup in the review and recosting, the company should abandon the project. The number of bad projects that make it to the operational stage serves as proof that their supporters often balk at this decision
>
> . . . . Senior managers need to create an environment that rewards honesty and courage and provides for more decision making on the part of project managers. Companies must have an atmosphere that encourages projects to succeed, but executives must allow them to fail.

The longer the project, the greater the necessity for the exit champions and project sponsors to make sure that the business plan has "exit ramps" such that the project can be terminated before massive resources are committed and consumed. Unfortunately, when a collective belief exists, exit ramps are purposefully omitted from the project and business plans. Another reason for having exit champions is so that the project closure process can occur as quickly as possible. As projects approach their completion, team members often have apprehension about their next assignment and try to stretch out the existing project until they are ready to leave. In this case, the role of the exit champion is to accelerate the closure process without impacting the integrity of the project.

Some organizations use members of a portfolio review board to function as exit champions. Portfolio review boards have the final say in project selection. They also have the final say as to whether or not a project should be terminated. Usually, one member of the board functions as the exit champion and makes the final presentation to the remainder of the board.

DISCUSSION QUESTIONS

The discussion questions are for classroom use to stimulate group thinking about PM 2.0. There are no right or wrong answers to most of the questions.

---

[11] D. Davis, "New Projects: Beware of False Economics," *Harvard Business Review*, March–April 1985, pp.100–101. Copyright © 1985 by the President and Fellows of Harvard College. All rights reserved.

1. What are the differences between a methodology and a project plan?
2. What are the advantages of having a methodology based upon rigid policies and procedures?
3. How often should a methodology undergo continuous improvements?
4. What are the differences between a design freeze and a specification freeze?
5. Who has the final say in the approval of a project's business case?
6. What factors can cause the project's assumptions to change over the project's life cycle?
7. What factors can cause the project's constraints to change over the project's life cycle?
8. Who is responsible for validating the project's assumptions?
9. Should metrics be established for the tracking of assumptions and constraints?
10. What should be a reasonable sequence of events for shutting down a project?

CHAPTER 7
PROJECT GOVERNANCE

7.0 INTRODUCTION

Historically, project managers interfaced with just one person acting as the project sponsor. The role of the sponsor was to assist the project manager with critical decision making and insulate the sponsor from politics and stakeholders. But with PM 2.0 and the increased complexity of projects, stakeholders are expected to take a more active role on projects. One person may not be able to make all necessary support decisions.

Project managers must now expect active involvement by stakeholders and other parties. The result will be risks in political interference in projects, stakeholder micromanagement, and possibly numerous changes in the direction of the project. It is wishful thinking to expect that this will not happen. As such, today's project managers using the concepts of PM 2.0 must have an understanding of governance, stakeholder relations management, and politics. Effective communications skills are a necessity.

7.1 NEED FOR GOVERNANCE

All projects have the potential of getting into trouble even in a PM 2.0 environment. But in general, project management can work well as long as the project's requirements are realistic and do not impose severe pressure upon the project manager and a project sponsor exists as an ally to assist the project manager when trouble does appear. In today's chaotic environment, this pressure appears to be increasing because:

- Companies are accepting greater risks and highly complex projects as a necessity for survival.
- Customers are demanding low-volume, high-quality products with some degree of customization.
- Project life cycles and the time for new product development are being compressed.

How important is stakeholder agreement?

- Enterprise environmental factors are having a greater impact on project execution.
- Customers and stakeholders want to be more actively involved in the execution of projects.
- Companies are developing strategic partnerships with suppliers, and each supplier can be at a different level of project management maturity.
- Global competition has forced companies to accept projects from customers that are all at a different level of project management maturity and with different reporting requirements.

These pressures are several of the reasons for the existence of PM 2.0 because they tend to slow down the decision-making processes at a time when stakeholders want the projects and processes to be accelerated. One person, while acting as the project sponsor, may have neither the time nor the capability to address all of these additional issues. Unless proper governance exists on the project, the result will be a project slowdown. The slowdown occurs because:

- The project manager may be expected to make decisions in areas where he or she has limited knowledge.
- The project manager may hesitate to accept full accountability and ownership for the projects.
- Excessive layers of management for reporting and decision making are being superimposed on top of the project management organization.
- Risk management is being pushed up to higher levels in the organization hierarchy to people with limited knowledge of risk management, resulting in delayed decisions.
- The project manager may demonstrate questionable leadership ability on some of the nontraditional or complex projects.

The problems resulting from these pressures may not be able to be resolved, at least easily and in a timely manner, by a single project sponsor. These problems can be resolved using effective project governance as long as members of the governance committee can come to an agreement in a timely manner.

7.2 DEFINING PROJECT GOVERNANCE

Project governance is actually a framework by which project decisions are made. Governance relates to decisions that define expectations, accountability, responsibility, the granting of power, or verifying performance. Governance relates to consistent management, cohesive policies, processes, and decision-making rights for a given area of responsibility. Governance enables efficient and effective decision making to take place.

Project management textbooks assert that project managers have single-person accountability to produce the assets or deliverables of the project. Accountability is usually defined as responsibility and commensurate authority. However, there are

How important is it for the governance committee to fully understand their roles and responsibilities?

Copyright © Scott Maxwell/Fotolia

degrees of both authority and responsibility. Project managers almost always have to share some degree of both authority and responsibility with those responsible for governance and the degree of sharing must be clearly defined at the onset of the project even though it may be subject to change throughout the project.

Every project can have different governance even if each project uses the same enterprise project management methodology. The governance function can operate as a separate process or as part of project management leadership. Governance is designed not to replace project decision making but to prevent undesirable decisions from being made.

Members of a governance committee can come to the table with their own agendas, even if the agendas are hidden from other committee members and the project manager. But the governance committee members must be in agreement as to their roles and responsibilities. If an understanding and agreement are not reached at the onset of the project, the project manager may not be able to get governance decisions made in a timely manner.

7.3 PROJECT VERSUS CORPORATE GOVERNANCE

Is project governance the same as corporate governance?

Copyright © Scott Maxwell/ LuMaxArt/Shutterstock

Project governance is not the same as corporate governance. Corporate governance consists of the set of processes, customs, policies, laws, and institutions affecting the way people direct, administer, or control a corporation. Corporate governance also includes the relationships among the many players involved (i.e., the stakeholders) and the alignment of the results to corporate goals. The principal players include the shareholders, management, and the board of directors. Other stakeholders include employees, suppliers, customers, banks and other lenders, regulators, the environment, and the community at large.[1] The people participating in corporate governance can be completely different from the people participating in project governance and the roles, responsibilities, and decision-making rights can be different as well. Corporate governance often follows the expression, "Rank has its privilege." As such, decisions may be dictated from the top down whereas in project governance decisions may follow a vote of the members of the committee regardless of the rank of the various committee members.

Governance on projects and programs sometimes fails because people confuse project governance with corporate governance. The result is that members of the committee are not sure what their role should be. Some of the major differences are:

Alignment: Corporate governance focuses on how well the portfolio of projects is aligned to and satisfies overall business objectives. Project governance focuses on ways to keep a project on track and to verify that value will be created at the project's completion.

Direction: Corporate governance provides strategic direction with a focus on how project success will satisfy corporate objectives. Project governance is more operational direction with decisions based upon the predefined parameters on project scope, time, cost, and functionality.

---

[1] Adapted from Wikipedia, the free encyclopedia.

Dashboards: Corporate governance dashboards are based upon financial, marketing, and sales metrics. Project governance dashboards have operations metrics on time, cost, scope, quality, action items, risks, and deliverables.

Membership: Corporate governance committees are composed of the seniormost levels of management. Project governance membership may include some membership from middle management.

Another reason why failure may occur is when members of the project or program governance group do not understand project or program management and the differences between them. This can lead to unwanted micromanagement by the governance committee. Some people believe that projects deliver value whereas programs create assets. Another difference is that project governance may be short lived.

7.4 ROLES, RESPONSIBILITIES, AND DECISION-MAKING AUTHORITY

Project governance is the management framework within which project decisions are made. Project governance is a critical element of any project while the accountabilities and responsibilities associated with an organization's business as usual activities are laid down in their organizational/corporate governance arrangements. Seldom does an equivalent framework exist to govern the development of its capital investments (i.e., projects). For instance, the organization chart provides a good indication of who in the organization is responsible for any particular operational activity the organization conducts. But unless an organization has specifically developed a project governance policy, no such chart is likely to exist for project development activity. Therefore, the role of project governance is to provide a decision-making framework that is logical, robust, and repeatable to govern an organization's capital investments. In this way, an organization will have a structured approach to conducting both its business-as-usual activities and its business change, or project, activities.[2]

There is always the question of what decisions must be made by the governance committee and what decisions the project manager can make. Ambiguous or overlapping roles and responsibilities lead to chaos. In general, the project manager should have the authority for decisions related to actions necessary to maintain the baselines. Governance committees must have the authority to approve scope changes above a certain dollar value and to make decisions necessary to align the project to corporate objectives and strategy. Tables 7-1 and 7-2 illustrate some of the possible differences in assignments between the project manager and the governance groups.

---

[2] Adapted from Wikipedia, the free encyclopedia.

TABLE 7-1 Difference in Responsibilities	
Governance Committee	**Project Managers**
Define expected final results	Define tactical plans
Define interim deliverables	Define resource needs
Define strategic targets	Look at resource availability
Define strategic objectives	Look at capacity planning
Define funding limits	Establish the baselines
Define environmental factors	Assess execution risks
Define executive involvement	Identify the metrics/KPIs
	Control of scope creep

TABLE 7-2 Differences in Decision-Making Authority	
Governance Committee	**Project Managers**
Strategic planning decisions	Preparation of the baselines
Changing the strategic objectives	Maintaining the baselines
Changing the business objectives	Negotiating for resources
Approval of scope changes	Limited mitigation of risks
Timing of the health checks	
Project cancellation	

7.5 GOVERNANCE FRAMEWORKS

Historically, governance was provided by a single project sponsor. This was particularly true for internal projects, especially those that were not too complex. Today, governance is a committee and can include representatives from each stakeholder's organization whether internal or external to the company. Table 7-3 shows various governance approaches based upon the type of project team. The membership of the committee can change from project to project and industry to industry. The membership may also vary based upon the number of stakeholders and whether the project is for an internal or external client. On long-term projects, membership can change throughout the project.

For projects external to the company, the number of governance committees can be significant. This is shown in Figure 7-1. For projects external to the organization the hierarchy of committees can change positions. There can also be significant conflicts and political issues between the members of various governance groups over the value of the project, scope changes, costs, and other such arguments. Stakeholder agreements between individuals and committees are difficult to achieve.

Can project governance work effectively in a virtual setting?

TABLE 7-3 **Project Governance Approaches**

Project Organizational Structure	Description	Governance Approach
Dispersed locally	Team members can be full time or part time. They are still attached administratively to their functional area.	Usually a single person is acting as the sponsor but may be an internal committee based upon the project's complexity.
Dispersed geographically	This is a virtual team. The project manager may never see some of the team members. Team members can be full time or part time.	Usually governance by committee and can include stakeholder membership. Virtual governance teams can and do work.
Colocated	All of the team members are physically located in close proximity to the project manager. The project manager does not have any responsibility for wage and salary administration.	Usually a single person acting as the sponsor.
Projectized	This is similar to a colocated team but the project manager generally functions as a line manager and may have wage and salary responsibilities.	May be governance by committee based upon the size of the project and the number of strategic partners.

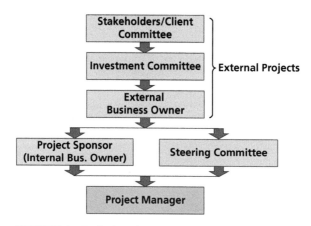

FIGURE 7-1 **Typical project governance structure**

7.6 THREE PILLARS OF PROJECT GOVERNANCE[3]

The three pillars of project governance are:

- Organizational structure
- People
- Information

---

[3] Adapted from Wikipedia, the free encyclopedia.

Structure refers to the governance committee structure. This can include several committees with varying degrees of responsibility and authority. The decision rights of all these committees and how they relate must be laid down in policy and procedural documentation. In this way, the project's governance can be integrated within the wider governance arena.

The effectiveness of the committee structure is dependent upon the people that populate the various governance committees. Committee membership is determined by the nature of the project; other factors come into play when determining membership of the various boards, which in turn determines which organizational roles should be represented on the committee.

The last pillar concerns the information that informs decision makers and consists of regular reports on the project, issues, and risks that have been escalated by the project manager and certain key documents that describe the project, foremost of which is the business case.

Core Project Governance Principles

Project governance frameworks should be based around a number of core principles in order to ensure their effectiveness.

Principle 1: Ensure a single point of accountability for the success of the project.

The most fundamental project accountability is accountability for the success of the project. A project without a clear understanding of who assumes accountability for its success has no clear leadership. With no clear accountability for project success, there is no one person driving the solution of the difficult issues that beset all projects at some point in their life. It also slows the project during the crucial project initiation phase since there is no one person to implement the important decisions necessary to place the project on a firm footing. The concept of a single point of accountability is the first principle of effective project governance.

However, it is not enough to nominate someone to be accountable—the right person must be made accountable. There are two aspects to this. The person who is accountable must hold sufficient authority within the organization to ensure they are empowered to make the decisions necessary for the project's success. Beyond this, however, is the fact that the right person from the correct area within the organization be held accountable. If the wrong person is selected, the project is no better placed than if no one was accountable for its success.

Principle 2: Project ownership is independent of asset ownership, service ownership, or other stakeholder group.

Often organizations promote the allocation of the project owner role to the service owner or asset owner with the goal of providing more certainty that the project will meet these owners' fundamental needs, which is also a critical project success measure. However, the result of this approach can involve wasteful scope inclusions and failure to achieve alternative stakeholder and customer requirements:

- The benefit of the doubt goes to the stakeholder allocated with the project owner responsibility, skewing the project outcome.

- Project owner requirements receive less scrutiny, reducing innovation and outcome efficiency.
- Different skill sets surround project ownership, asset ownership, and service ownership, placing sound project decision making and procedure at risk.
- Operational needs always prevail, placing the project at risk of being neglected during such times.
- Project contingencies are at risk of being allocated to additional scope for the stakeholder allocated project ownership.

The only proven mechanism for ensuring that projects meet customer and stakeholder needs while optimizing value for money is to allocate project ownership to a specialist party who otherwise would not be a stakeholder to the project.

Principle 3: Ensure separation of stakeholder management and project decision-making activities.

The decision-making effectiveness of a committee can be thought of as being inversely proportional to its size. Not only can large committees fail to make timely decisions, but also those they do make are often ill considered because of the particular group dynamics at play.

As project decision-making forums grow in size, they tend to morph into stakeholder management groups. When numbers increase, the detailed understanding of each attendee of the critical project issues reduces. Many of those present attend not to make decisions but as a way of finding out what is happening on the project. Not only is there insufficient time for each person to make their point, but also those with the most valid input must compete for time and influence with those with only a peripheral involvement in the project. Further not all present will have the same level of understanding of the issues and so time is wasted bringing everyone up to speed on the particular issues being discussed. Hence, to all intents and purposes, large project committees are constituted more as a stakeholder management forum than a project decision-making forum. This is a major issue when the project is depending upon the committee to make timely decisions.

There is no question that activities such as project decision making and stakeholder management are essential to the success of the project. The issue is that they are two separate activities and need to be treated as such. This is the third principle of effective project governance. If this separation can be achieved, it will avoid clogging the decision-making forum with numerous stakeholders by constraining its membership to only those select stakeholders absolutely central to its success.

There is always the concern that this solution will lead to a further problem if disgruntled stakeholders do not believe that their needs are being met. Whatever stakeholder management mechanism that is put in place must adequately address the needs of all project stakeholders. It will need to capture their input and views and address their concerns to their satisfaction.

Principle 4: Ensure separation of project governance and organizational governance structures.

Project governance structures are established precisely because it is recognized that organization structures do not provide the necessary framework to deliver a project. Projects require flexibility and speed of decision making and the hierarchical mechanisms associated with organization charts do not enable this. Project governance structures overcome this by drawing the key decision makers out of the organization structure and placing them in a forum, thereby avoiding the serial decision-making process associated with hierarchies.

Consequently, the project governance framework established for a project should remain separate from the organization structure. It is recognized that the organization has valid requirements in terms of reporting and stakeholder involvement. However, dedicated reporting mechanisms established by the project can address the former and the project governance framework must itself address the latter. What should be avoided is the situation where the decisions of the steering committee or project board are required to be ratified by one or more persons in the organization outside of that project decision-making forum.

Adoption of this principle will minimize multilayered decision making and the time delays and inefficiencies associated with it. It will ensure a project decision-making body empowered to make decisions in a timely manner.

Additional and complementary principles of governance also exist:

- The board has overall responsibility for governance of project management.
- The roles, responsibilities, and performance criteria for the governance of project management are clearly defined.
- Disciplined governance arrangements supported by appropriate methods and controls are applied throughout the project life cycle.
- A coherent and supportive relationship is demonstrated between the overall business strategy and the project portfolio.
- All projects have an approved plan containing authorization points, at which the business case is reviewed and approved. Decisions made at authorization points are recorded and communicated.
- Members of delegated authorization bodies have sufficient representation, competence, authority, and resources to enable them to make appropriate decisions.
- The project business case is supported by relevant and realistic information that provides a reliable basis for making authorization decisions.
- The board or its delegated agents decide when independent scrutiny of projects and project management systems is required and implement such scrutiny accordingly.
- There are clearly defined criteria for reporting project status and for the escalation of risks and issues to the levels required by the organization.
- The organization fosters a culture of improvement and of frank internal disclosure of project information.
- Project stakeholders are engaged at a level that is commensurate with their importance to the organization and in a manner that fosters trust.

Additional principles exist where projects are multiowned. *Multiowned* is defined as being a project where the board shares ultimate control with other parties. The principles are:

- There should be formally agreed-upon governance arrangements.
- There should be a single point of decision making for the project; refer to principle 2.

- There should be a clear and unambiguous allocation of authority for representing the project in contacts with owners, stakeholders, and third parties.
- The project business case should include agreed-upon and current definitions of project objectives, the role of each owner, their incentives, inputs, authority, and responsibility.
- Each owner should assure itself that the legal competence and obligations and internal governance arrangements of coowners are compatible with its acceptable standards of governance for the project.
- There should be project authorization points and limiting constraints to give owners the necessary degree of control over the project.
- There should be agreed-upon recognition and allocation or sharing of rewards and risks taking into account ability to influence the outcome and creating incentives to foster cooperative behavior.
- Project leadership should exploit synergies arising from multiownership and should actively manage potential sources of conflict or inefficiency.
- There should be a formal agreement that defines the process to be invoked and the consequences for assets and owners when a material change of ownership is considered.
- Reporting during both the project and the realization of benefits should provide honest, timely, realistic, and relevant data on progress, achievements, forecasts, and risks to the extent required for good governance by owners.
- There should be a mechanism in place to invoke independent review or scrutiny when it is in the legitimate interests of one or more of the project owners.
- There should be a dispute resolution process agreed upon between owners that does not endanger the achievement of project objectives.

A key role in project governance is that of the project sponsor. The project sponsor has three main areas of responsibility: to the board, to the project manager, and to project stakeholders.

For the board, the sponsor provides leadership on culture and values, owns the business case, keeps the project aligned with the organization's strategy and portfolio direction, governs project risk, works with other sponsors, focuses on realization of benefits, recommends opportunities to optimize cost/benefits, ensures continuity of sponsorship, provides assurance, and provides feedback and lessons learnt.

For the project manager, the sponsor provides timely decisions, clarifies the decision-making framework, clarifies business priorities and strategy, communicates business issues, provides resources, engenders trust, manages relationships, supports the project manager's role, and promotes ethical working.

For other project stakeholders, the project sponsor engages stakeholders, governs stakeholder communications, directs client relationship, directs governance of users, directs governance of suppliers, and arbitrates between stakeholders.

Project governance will:

- Outline the relationships between all internal and external groups involved in the project
- Describe the proper flow of information regarding the project to all stakeholders

- Ensure the appropriate review of issues encountered within each project
- Ensure that required approvals and direction for the project is obtained at each appropriate stage of the project

Important specific elements of good project governance include:

- A compelling business case, stating the objects of the project and specifying the in-scope and out-of-scope aspects
- A mechanism to assess the compliance of the completed project to its original objectives
- Identifying all stakeholders with an interest in the project
- A defined method of communication to each stakeholder
- A set of business-level requirements as agreed to by all stakeholders
- An agreed-upon specification for the project deliverables
- The appointment of a project manager
- Clear assignment of project roles and responsibilities
- A current, published project plan that spans all project stages from project initiation through development to the transition to operations
- A system of accurate upward status and progress reporting, including time records
- A central document repository for the project
- A centrally held glossary of project terms
- A process for the management and resolution of issues that arise during the project
- A process for the recording and communication of risks identified during the project
- A standard for quality review of the key governance documents and of the project deliverables

In summary and using IT as an example, governance primarily deals with connections between business focus and IT management. The goal of clear governance is to ensure the investment in IT generates business value and mitigates the risks that are associated with IT projects.

7.7 MISINTERPRETATION OF INFORMATION

In an ideal situation, all members of the governance committee will see the same dashboard. The governance committee members will be trained in how to evaluate the information and all of the members will be in agreement as to the interpretation of the information. Unfortunately, there are several reasons why stakeholder disagreements occur:

- Stakeholders bring to the table their own agendas, which may be hidden and may not be aligned with the strategic objectives of the project.
- Stakeholders may have long- or short-term interests which are not supported by the metric information.
- Each of the stakeholders may have requested additional metrics which may provide conflicting information compared to the common metrics.
- Stakeholders may not believe that the metric information is real.

When disagreements occur, the following may occur:

- Governance committee member conflicts will occur.
- The committee will make seat-of-the-pants rather than informed decisions.
- Decisions may be delayed, thus impacting the performance of the project.
- The committee may usurp the authority of the project manager.
- The committee may try to change some of its roles and responsibilities.
- The committee may request additional information, dashboards, or additional text material.
- Members of the committee may try to pull rank on one another.
- The committee member providing the greatest financial support for the project may withdraw the support.
- Committee members may resign from the committee, thus slowing down the decision-making process.

What happens if stakeholders disagree on the meaning of the metric information?

7.8 FILTERING THE INFORMATION

We all know of case studies where the project team misrepresented the information about the project's status to make the project look better than it actually was. Hopefully, the governance committee can differentiate fact from fantasy. But what happens when members of the governance committee want to see the information either in raw data format or before it is officially published in a report or presented on a dashboard?

We must remember that members of the governance committee may have hidden agendas and want the information to appear as though everything is going along well. The information presented to the governance committee members may not stop at the committee level. The information may then be transmitted from the committee members to their superiors. Based upon the timing of the information, especially if the information is not very favorable, the committee members could find that this information may have an impact on their career, salary increases, bonuses, or political interests. Most project managers have no idea where the information goes once it reaches the governance committee. In such situations, the governance committee may ask the project manager to stretch the truth or the committee members themselves may alter the information in the dashboard prior to transmitting it elsewhere.

From a project manager's perspective, honesty is the best policy. Stretching the truth will eventually come back and haunt the project manager once the situation is revealed.

What happens if the governance committee wants to change the dashboard information?

7.9 UNDERSTANDING POLITICS IN PROJECT ENVIRONMENT

The completion of a project requires three P's: people, processes, and politics. People and politics have a strong correlation. Simply assigning people to the project does not necessarily mean that they will always make decisions that are in the best interest of the project. When people are first assigned to a new project, they ask themselves, "What's in it for me? How will my career benefit from this assignment?" This is the onset of project politics and can lead to the filtering of information described in the previous section.

This type of thinking can permeate all levels of management on a project, including those responsible for the governance of the project. People tend to play politics to get what they want and this gamesmanship creates barriers that the project manager must overcome. People are motivated by the rewards they can receive from the formal structure of the company and also from the informal political power structure that exists. Barriers are created when an individual's rewards from either structure are threatened. The barriers lead to conflicts and can involve how the project will be planned, who will be assigned to specific activities, especially those activities that may receive high-level visibility, which approach to take to solve a problem, and other such items that are often hidden agenda items. Some people may even want to see the project fail if it benefits them.

Political savvy is an essential skill for today's project manager. One can no longer rely solely upon technical or managerial competence when managing a project. You must understand the political nature of the people and organizations you must deal with. You must understand that politics and conflicts are inevitable and are a way of life in project management. Project managers of the future must become politically astute. Unfortunately, even though there are some books published on politics in project management,[4] there has been limited research conducted on project management politics compared to other areas of the **PMBOK® Guide**.

Political Risks

On large and complex projects, politics is often treated as a political risk, especially when the project is being conducted in the host's country and subjected to government interference or political violence. The factors often considered as part of political risks include:

- Political change such as a new party elected into power
- Changes in the host country's fiscal policy, procurement policy, and labor policy
- Nationalization or unlawful seizure of project assets and/or intellectual property
- Civil unrest resulting from a coup, acts of terrorism, kidnapping, ransom, assassinations, civil war, and insurrection
- Significant inflation rate changes resulting in unfavorable monetary conversion policies
- Contract failure such as license cancellation and payment failure

Many of these risks emanate from an understanding of the enterprise environmental factors. Under normal situations, the enterprise environmental factors are monitored by the project sponsor or the governance committee, and they in turn may provide assistance in managing political risks. But when the project is being conducted within the host's country and where the project manager may not be in direct contact with the governance committee, it is usually the project manager who has to deal with the political risks.

---

[4] See J. K. Pinto, *Power & Politics in Project Management*, The Project Management Institute, Newtown Square, PA, 1996, and B. Irwin, *Managing Politics and Conflicts in Projects*, Management Concepts, Vienna, VA, 2008.

The larger and more complex the project, the larger the cost overrun. And the larger the cost overrun, the greater the likelihood of political intervention. In some countries, such as in the United States, escalating problems upward usually implies that the problem ends up in the hands of the project sponsor. But in other countries, especially emerging market nations, problems may rise beyond the project's governance committee and involve high-level government officials. This is particularly true for megaprojects that are susceptible to large cost overruns.

Reasons for Playing Politics

There are numerous reasons why people play political games. Some of the common reasons are:

- Wanting to maintain control over scarce resources
- Seeking rewards, power, or recognition
- Maintaining one's image and personal values
- Having hidden agendas
- Fear of the unknown
- Control over who gets to travel to exotic locations
- Control over important information since information is a source of power
- Getting others to do one's work
- Seeing only what one wants to see
- Refusing to accept or admit defeat or failure
- Viewing bad news as a personal failure
- Fearful of exposing mistakes to others
- Viewing failure as a sign of weakness
- Viewing failure as damage to one's reputation
- Viewing failure as damage to one's career

All of these are reasons that may benefit you personally. There are also negative politics where political games are played with the intent of hurting others, which in turn may end up benefiting you personally. Some examples would be:

- Wanting to see the project fail
- Fearful of unwanted change if the project succeeds
- Wanting to damage someone else's image or reputation, especially if they stand in the way of your career advancement
- Berating the ideas of others to strengthen your position

Situations Where Political Games Will Occur

While politics can exist on any project and during any project life-cycle phase, there are some specific situations where history has shown us that politics are most likely to occur:

- Trying to achieve project management maturity within a conservative culture
- During mergers and acquisitions where the "landlord" and the "tenant" are at different levels of project management maturity

- Trying to get an entire organization to accept a project management methodology that was created by one functional area rather than a committee composed of members from all functional areas (i.e., the not-invented-here syndrome)
- Not believing that the project can be completed successfully and wanting to protect oneself
- Having to change one's work habits and do things differently if the project is a success
- When problems occur, not knowing where they will end up for resolution
- Believing that virtual teams are insulated from project politics
- The larger and more complex the project, the greater the chances of political interference
- The larger the size of the governance committee, the greater the chance for disagreements and political issues to appear
- Failing to understand effective stakeholder relations management practices
- The more powerful the people are on the project, the greater the chance that they will be involved in project politics
- Employees who are recognized as prima donnas are more prone to play political games than the average worker
- The project is in trouble and trade-offs are needed to recover a possibly failing project.

Governance Committee

Project politics usually ends up pushing the project in a direction different from the original SOW. The push can originate within your own senior management, some of your project team members, the customer, and even some of the stakeholders. Each may want a slightly different project outcome and your job is to try to find a way to appease everyone.

On the surface, the simplest solution appears to be the creation of a governance committee composed of senior managers from your company, representation from the customer's company, and representatives from various stakeholder groups. Now, it seems that you can let the governance committee resolve all of the conflicts among themselves and give you a unified direction for the project. Gaining support from a higher power certainly seems like the right thing to do. Unfortunately, there is still the possibility that the committee cannot come to an agreement, and even if they appear to be in agreement, certain members of the committee may still try to play politics behind the scenes. The existence of the governance committee does not eliminate the existence of project politics. People that serve on a governance committee often play the political game in order to enhance their power base.

Most companies have limited funds available for projects. The result is an executive-level competition for project funding that may serve the best interest of one functional area but not necessarily the best interest of the entire company. Executives may play political games to get their projects approved ahead of everyone else, viewing this as an increase to their power base. But the governance committee may include executives from those functional areas that lost out in the battle for project funding, and these executives may try to exert negative political influence on the project even so far as to hope that the project would fail. The result often occurs when a project manager is assigned to such a project and brought on board after the project is approved, never fully understanding until well into the project the politics that were played during project approval and initiation.

Friends and Foes

It is often difficult to identify quickly which people are friends or foes. Not all people that have political agendas are enemies. Some people may be playing the political game for your best interest. It is therefore beneficial to identify if possible from the personal agendas that people have whether they are friends or foes. This implies that you must communicate with them, perhaps more informally than formally, to understand their agendas. Reading body language is often a good way to make a first guess if someone is a friend or foe.

One possible way to classify people might be:

True Supporters: These are people that openly demonstrate their willingness to support you and your position on the project.

Fence-Sitters: These are people that you believe will support you down the road as long as you prove to them that you are deserving of their trust and support. You may need to spend extra time with them to show them your position and to gain their support.

True Unknowns: Unlike fence-sitters who may be won over to your way of thinking, these people are true unknowns. They may have hidden agendas that are not in your best interest, but they are relatively quiet and may have not yet expressed their concerns. These people could pose a serious threat if they are adamantly opposed to the direction in which the project is proceeding.

True Enemies: These are people that have made it quite clear that they are unlikely to support your views. You understand their position and probably are quite sure how they will respond to you and the direction the project is taking.

Attack or Retreat

When people play political games on projects, there are two facts that we seem to take for granted. First, these people are most likely experienced in playing such games, and second they expect to win. Based upon whom the conflict is with, you must then decide whether to aggressively attack them or retreat. Simply taking no action is a form of withdrawal and you are sure to lose the battle.

The first rule in battle is to gather as much intelligence as you can about your enemy. As an example, as part of stakeholder relations management, we can map project stakeholders according to Figure 7-2. Stakeholder mapping is most frequently displayed on a grid comparing their power and their level of interest in the project.

Manage Closely: These are high-power, interested people that can make or break your project. You must put forth the greatest effort to satisfy them. Be aware that there are factors that can cause them to change quadrants rapidly.

Keep Satisfied: These are high-power, less interested people who can also make or break your project. You must put forth some effort to satisfy them but not with excessive detail that can lead to boredom and total disinterest. They may not get involved until the end of the project approaches.

Keep Informed: These are people with limited power but keenly interested in the project. They can function as an early warning system of approaching problems and may be technically astute to assist with some technical issues. These are the stakeholders who often provide hidden opportunities.

Monitor Only: These are people with limited power and may not be interested in the project unless a disaster occurs. Provide them with some information but not with too much detail such that they will become disinterested or bored.

When you go on offense and attack the people playing politics, you must have not only ammunition but also backup support if necessary. You must be prepared to show how the political decision might impact the constraints on the project as well as the accompanying baselines. Based upon the power and influence level of your opponent according to Figure 7-2, you may need other stakeholders to help you plead your case. It is highly beneficial to have supporters at the same level of position power or higher than the people playing the political game.

Not all political battles need to be won. People who play politics and possess a great deal of power may also have the authority to cancel the project or assist in the recovery process. In such cases where people possess the power to cancel the project, retreating from a political battle may be the only viable option. If you truly alienate the people playing power games, the situation can deteriorate even further. There is always the chance that you may have to work with the same people in the future. In any case, the best approach is to try to understand the people playing politics, the reason why they are playing politics, and how much power and influence they have over the final decision.

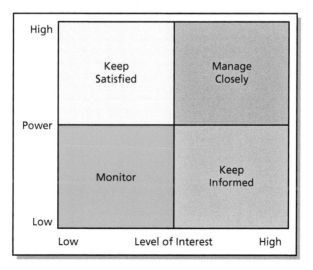

FIGURE 7-2 Stakeholder mapping

Need for Effective Communications

While it is not always possible to tell when someone is playing or intends to play the political game on your project, there are some tell-tale signs that this may be happening. Some of the signs include:

- People do not care about your feelings.
- People avoid discussing critical issues.
- People never ask you about your feelings on the matter.
- People procrastinate making decisions.
- People have excuses for not completing action items.
- People discuss only those items that may benefit them personally.

While project managers may not have any control over these tell-tale signs, project managers can make the situation worse through ineffective communications. To minimize the political impact on a project, the project manager should consider using the following practices:

- Listen carefully before speaking and do not jump to conclusions.
- Make sure you understand what others are saying and try to see the issue from their point of view.
- All informal communications should be followed up with a memo outlining what was discussed and to make sure that there were no misunderstandings.
- Before stating your point of view, make sure that you have gathered all of the necessary supporting information.
- Make sure that you have a clear understanding of how culture impacts the way that people are communicating with you.
- If you must provide criticism, make sure that it is constructive rather than personal criticism.
- When resolving political issues, there will be winners and losers. It is not a matter of just picking a winner. You must also explain to everyone why you selected this approach and likewise why the other approaches were not considered. This must be done tactfully.
- If the situation cannot be managed effectively, do not be embarrassed to ask senior management for advice and assistance.
- Ineffective communications encourages lying, which in turn generates additional political games to be played accompanied by a great deal of mistrust.

Project managers must be careful when discussing politics with team members, the client, and stakeholders. The information could be misunderstood or filtered, especially if people hear what they want to hear. The result could be additional politics that were unexpected, and friends could easily turn into foes.

Power and Influence

Effective communication skills alone cannot resolve all political situations. To understand why, we must look at how project management generally works. If all projects stayed within the traditional hierarchy, someone would have the ultimate authority to resolve political issues. But since most projects are managed outside the traditional hierarchy, the burden for the resolution of conflicts and political issues usually falls upon the shoulders of the project manager even if a governance committee is in place. The governance committee may very well be the cause of the conflict.

On the surface, it seems like the simplest solution would be to give the project manager sufficient authority to resolve political issues. But projects are usually executed outside of the traditional hierarchy, thus limiting the authority that the project manager will possess. This lack of formal authority makes the project manager's job difficult. While project charters do give project managers some degree of authority for a given project, most project managers still have limitations because:

- The project managers must negotiate with functional managers for qualified resources.
- The project managers may not be able to remove employees from a project without the functional manager's concurrence.
- The project managers generally have no direct responsibility for wage and salary administration.
- The project managers may possess virtually no reward or punishment power.
- If employees are assigned to multiple projects, the project managers may not be able to force the employees to work on their projects in a timely manner.

With a lack of position power which comes from the traditional hierarchy and without the ability to reward or punish, the project manager must rely upon other forms of power and the ability to influence people. Behavioral skills such as effective communications, motivation techniques, conflict management, bargaining, and negotiations are essential to resolve political disputes. Unfortunately most project managers lack political savvy and have poor conflict resolution skills.

Managing Project Politics

While project politics are inevitable, there are actions the project manager can take to minimize or control political issues. Some of these actions are:

- Gather as much information as you can about the political situation.
- Make sure that everyone fully understands the impact of the political situation upon the project's baselines.
- Try to see the picture through the eyes of the person playing politics.
- Try to form a coalition with the people playing politics.
- See if your sponsor or the governance committee can insulate you from the political games.
- Having a structured decision-making process as part of your project management methodology can reduce some of the political games.
- Try to determine one's political position by reading their body language.
- If the political situation cannot be resolved quickly, demonstrate a willingness to compromise as long as the integrity of the project is not sacrificed.

Power breeds politics and politics in turn breeds power. Expecting to manage a project without any political interference is wishful thinking rather than reality.

We cannot predict customer and stakeholder behavior. Sometimes the political situation occurs without any early warning signs.

Nobody can agree on a definition of organizational or project politics. Politics can appear in many shapes, forms, and sizes. Therefore the project manager must develop superior behavioral skills to deal with political situations. The danger in not being able to manage political situations correctly is redirection or misdirection of the project.

7.10 MANAGING GLOBAL STAKEHOLDER RELATIONS

We mentioned previously the necessity to perform stakeholder mapping with regard to their power and influence. This assumes of course that we know who the stakeholders are and understand their concerns. This information is usually known with regard to internal stakeholders. However, when dealing with external stakeholders on global projects, there are significant unknowns that can influence the direction of the project. To make matters worse, global projects are implemented in complex institutions that are highly impacted by a sociopolitical environment. Some of the critical factors when dealing with global stakeholders are:

Can we work with global stakeholders the same way as non-global stakeholders?

Copyright © Scott Maxwell/
LuMaxArt/Shutterstock

- Not all of the stakeholders are easily identifiable.
- There are significantly more stakeholders than on nonglobal projects.
- Global stakeholders are more likely to change over the duration of the project.
- There may be hidden stakeholders higher up in the organizational hierarchy than the project's global stakeholders.
- Not all of the issues facing the stakeholders are easy identified.
- Solutions to problems on the project may need to be resolved higher up than the global stakeholders.
- Not all issues with global stakeholders can be resolved the same way we resolve issues with nonglobal stakeholders.
- Global stakeholders may have more hidden agendas than nonglobal stakeholders.
- Conflicts with global stakeholders may need to be approached with a different set of conflict resolution modes than we use on traditional projects.

Aaltonen and Sivonen have written an excellent paper discussing conflict resolution modes for dealing with stakeholder pressure on global projects.[5] In their paper, they discuss five conflict resolution strategies:

- Adaptation strategy: obeying stakeholder rules, policies, and procedures
- Compromising strategy: negotiation and dialogue
- Avoidance strategy: loosening attachments to the stakeholders
- Dismissal strategy: ignoring their demands
- Influence strategy: shaping proactively the demands of the stakeholders

Emerging market global stakeholders can respond differently to issues than global stakeholders in other markets. It is of critical importance to understand the operating and cultural environment in which the global stakeholders reside.

---

[5] K. Aaltonen and R. Sivonen, Response Strategies to Stakeholder Pressure in Global Projects, *International Journal of Project Management*, 27 (2009), 131–141.

7.11 FAILURE OF PROJECT GOVERNANCE

Simply placing an executive on a project governance committee is no guarantee that the correct decisions will be made. Executives are prone to making mistakes just like everyone else. The most common mistake made by executives on governance committees is when they say that they do not want to hear any bad news. This can lead to the filtering of information. Even though it is a thin line between governance failures and project failures, poor decisions by a governance committee can be more costly than poor decisions made by a project manager.

In addition to the mistakes discussed in the previous sections, other significant mistakes include:

Lessons Learned Are Not Captured: We tend to capture lessons learned and best practices from the project team but not from the governance committee. Best practices can be learned from both successes and failures. Members of governance committees are reluctant to admit mistakes that can go into lessons learned files, identify governance shortfalls, and be read by other stakeholders now or in the future. The end result is that we have very little lessons learned data from governance failures.

Refusal to Listen to Concerns of the Project Manager: Not all governance members have ever managed projects or attended project management training courses. The argument they use is that they are too busy to attend even a one-day class. As such, they often lack the knowledge of what it takes to make a project successful. The result is that they refuse to listen to recommendations of the project managers and often replace projects managers for not agreeing with governance decisions. It is an uphill battle for project managers to influence the decisions made by a governance committee even if the project manager's recommendations are based upon sound facts.

Making Political Decisions: Governance team members often regard political decisions as being more important than sound project decisions. When this occurs, the impact on the project can be devastating and lead to project failure. Political decisions can include personal gain such as from bonuses and promotions. Politics can force greed to win out over integrity. It is not always possible to separate the success of a project from the success of people's careers.

Changing Stakeholders: When new stakeholders come on board an ongoing project, they must be willing to abide by the agreements of the previous stakeholders unless there is a significant change needed for the direction of the project. The worst situation occurs when new stakeholders join the governance committee and have political agendas that may be detrimental to the project.

Refusing to Pull Plug: Governance team members must be willing to cancel projects that can no longer be successful. Projects should be canceled before they become total disasters. Unfortunately, politics, image, reputation, and fear of retribution cause executives to let the project continue to completion while they look for someone to blame. Canceling projects that are in the public media are often seen as weakening one's position. Unpopular decisions seem to affect executives more so than project managers.

Rank Has Its Privilege: In some environments, members of the project governance committee believe that the project team members are so far beneath them in rank that they refuse to communicate directly with the project team. This could be the result of a cultural issue. The impact is usually a breakdown in communications and filtering of important information.

Excessive Governance: If governance team members view the success of the project as an opportunity for advancement and recognition, the result could be excessive governance. This usually leads to micromanagement by the governance committee.

Conflicting Visions: The governance committee and the project manager must have the same vision about the project. This does not imply that the governance committee will always agree with the decisions made by the project manager. Even with a clear vision, there will be a balance between being completely open and sharing all information and being secretive. The governance committee may withhold the releasing of sensitive information, especially if it is business-related information as opposed to project-based information.

Everything Is Top Priority: Governance committees need to understand that not all projects are the top priority. While governance team members that are assigned to multiple projects may recognize this fact, many others do not and reassign critical resources from important projects to satisfy their own pet project.

Not Validating Business Case: Members of the governance committee may not have been involved in the development of the business case. However, they should periodically revalidate that the business case is still correct.

7.12 SAVING DISTRESSED PROJECTS

Projects can and do get into trouble. Sometimes, despite the efforts of the project manager, the safety net for the project rests in the hands of the governance committee.

We train project managers in recovery project management practices. The project manager then decides what trade-offs can be made to salvage whatever value exists on the project. However, it is then the decision of the governance committee as to whether a safety net should be given to the project manager. There are situations where it may be better to euthanize the project than to recover only a portion of the value. As expected, decisions to provide a safety net may be based upon each committee member's personal agenda.

Need for a safety net

DISCUSSION QUESTIONS

The discussion questions are for classroom use to stimulate group thinking about PM 2.0. There are no right or wrong answers to most of the questions.

1. What are the differences between sponsorship and governance?
2. Which would have the greater likelihood of causing damage to a project: sponsorship or governance?
3. What should be the maximum size of a governance committee?
4. What factors could cause a change in sponsorship over the life of a project?
5. What factors could cause a change in the membership of a governance committee over the life of a project?
6. Can a governance committee perform virtually?
7. How do we make sure that governance committee members do not misinterpret the information provided to them on a dashboard?
8. Who decides upon the roles and responsibilities for the governance committee and the project team assuming they both cannot have the same roles and responsibilities?
9. What decisions can a governance committee make that would jeopardize the health of a project?
10. Can a governance committee assist the project team in recovery of a failing project? If so, how?

**What is the role of
the project manager
during portfolio
selection activities?**

CHAPTER 8

ROLE OF PROJECT MANAGER IN STRATEGIC PLANNING AND PORTFOLIO MANAGEMENT

8.0 INTRODUCTION

After more than five decades, project management has matured from what was once considered to be just a fad that would soon disappear into a strategic competency and career path necessary for the growth and survival of the firm. Project management is now being used in virtually every industry and in all parts of the business, including project selection activities. Project managers are now considered business people rather than just project managers.

Today, project management is recognized as a series of processes that can be used on every project, regardless of its length or complexity, the project's dollar value, or the project's exposure to risk. Yet the parts of the business where project management has been slow in being accepted, at least up to now, has been in strategic planning and portfolio management. Projects critical to strategic planning were normally managed by functional managers rather than project managers. All of that has changed with PM 2.0.

One can always argue that managing strategic planning execution projects is no different than managing any other type of project. While this argument may have merit in some companies, there are several important differences that must be considered. Specifically, project managers must think strategically rather than tactically or operationally, and they may have to change from traditional project management leadership to strategic leadership based upon the complexity of the project. PM 2.0 recognizes that project managers are managing both strategic and tactical projects and that the leadership style of the project managers may have to change.

8.1 WHY STRATEGIC PLANS OFTEN FAIL

To understand how project management can benefit strategic planning, it is important to understand why some strategic plans fail. Historically, project managers never participated in project selection activities because they most often came from the technical ranks of the company. Projects that were handled by project managers were mostly tactical rather than strategic projects.

Strategic projects were managed by functional managers. When functional managers receive year-end bonuses based upon yearly profitability, they tend to retain their best resources for the short-term projects that affect yearly profitability. In other words, a five-year strategic project may not get the best resources until perhaps the fifth year of execution. To solve this problem, management now believes that project managers are qualified to manage these strategic projects.

Many functional managers never fully understood project management and yet were assigned to manage strategic projects. Some of the other more common reasons for failure that resulted from either a poor understanding or application of project management concepts are:

- Neglecting to understand how the enterprise environmental factors can influence senior management's vision of the future
- Inadequate understanding of consumer behavior or the client's actions
- Improper research prior to project approval
- Poorly defined or ill-defined scope
- Poorly documented business case resulting in the approval of the wrong project
- Failing to get executive and stakeholder buy-in right from the start
- Poor executive governance once the strategy begins to be implemented
- Constantly changing the membership of the governance team
- Overestimating resource competencies needed for project execution
- Poor capacity-planning efforts resulting in understaffed projects
- Functional managers refusing to commit the proper resources for the duration of the strategic project
- Failing to get employee commitment to the project
- Failing to explain the project well to the project execution team
- Failing to explain to the execution team the incentives or financial benefits of working on this long-term project
- Failing to understand the magnitude of the organizational change needed for the project to be a success
- Unable to manage change effectively
- Failing to consider the impact of changes in technology during the execution of the project
- Poor estimating of time and cost
- Having an execution team that is unable to work with ill-defined or constantly changing requirement
- Poor integration of the project across the entire organization
- Inadequate communications

There are numerous other reasons for the failure of strategic planning execution projects. The above causes could happen on any project, but on strategic planning execution projects the potential damage to the firm could be quite severe.

8.2 PROJECT MANAGEMENT: EXECUTIVE PERSPECTIVE

With the ability to produce repeated successes on projects, it is no wonder that executives are now realizing the value in using project management for the execution of a strategic plan. There are several reasons why executives see value in using project management for these activities:

- Execution takes significantly more time than planning and consumes more resources. Executives do not have the time to spend possibly years coordinating and integrating work across a multitude of functional areas.
- Without a successful implementation plan, strategic planning cannot succeed.
- Project managers can successfully manage the dysfunctional separation between planning and execution.
- Long-term strategic objectives must be broken down into short-term objectives to simplify execution. This can be done easily using project management tools and a WBS.
- Project management staffing techniques, possibly with the use of a PMO, can match the proper resources to the projects.
- The organizational process assets used in project management, including dashboard reporting systems, can keep senior management updated on project status on a daily basis.
- Strategic planning objectives, because of the long time duration, are highly organic and subject to change. Project managers know how to manage and control change.

8.3 STRATEGIC PLANNING: PROJECT MANAGEMENT PERSPECTIVE

Strategic planning is an organization's process of defining where and how it would like to be positioned in the future. The future may be measured in a 3-year, 5-year, or 10-year (or longer) window. The strategic plan is based upon the firm's vision, mission, social consciousness, and values. Strategic planning requires an understanding of the firm and its environment. Executives, more so than project managers, have a better understanding of the enterprise environmental factors, namely products offered, markets served, present and future technologies, supplier base, labor markets, economic conditions, the political environment, and regulatory requirements. But if project managers are expected to meet these needs, then the project managers must be strategically oriented rather than just tactically oriented.

Executives establish high-level objectives for *what they want done*. Often, this is nothing more than a wish list that may or may not border on reality. The role of the project manager is to determine *if it can be done*. This requires a clear business case for each project, a scope statement, and use of the WBS to break down the high-level objectives into subobjectives or lower level objectives that are easier to understand and accomplish. If the project manager and the project team believe that is can be done, then a formalized project action plan is created. According to Wikipedia, the free encyclopedia:

> One of the core goals when drafting a strategic plan is to develop it in a way that is easily translatable into action plans. Most strategic plans address high-level initiatives and over-arching goals, but don't get articulated (translated) into day-to-day projects and tasks that will be required to achieve the plan. Terminology or word choice, as well as the level at which a plan is written, are both examples of easy ways to fail at translating your strategic plan in a way that makes sense and is executable to others. Often, plans are filled with con-ceptual terms which don't tie into day-to-day realities for the staff [project team] expected to carry out the plan.

On the surface, it may appear that strategic planning execution projects can be treated as any other project. However, if we look at the areas of knowledge in the **PM-BOK® Guide**, we can see some significant differences mostly attributed to the length of the project. A few of these differences are shown in Table 8-1.

TABLE 8-1 *PMBOK® Guide* and Execution of Strategic Projects	
Area of Knowledge	**Strategic Planning Project Impacts**
Integration management	The integration of the effort may very well span the entire organization both do-mestically and globally.
Scope management	The scope can change as technology changes. The length of the project makes it imperative that an effective scope change control process exists. The scope base-line may appear as a moving window requiring constant updates.
Time management	Matching the right people and their availability to the constantly changing scope will play havoc with scheduling. Losing people due to fire fighting in the functional areas may have a serious impact.
Cost management	Predicting the true cost of the project will be almost impossible. Reestimation must take place on a routine basis to make sure that the benefits and business value still exceed the cost.
Quality management	Customers' expectations of quality and competitive forces can cause major changes to the direction of the project.
Human resource management	The longer the project, the greater the likelihood that changes in resources will occur, possibly for the worse. Long-term motivation may be difficult to do.
Communication management	Communication requirements can span the entire company. Changes in stake-holders will also have a serious impact on the communication plan.
Risk management	The project may be required to have a dedicated risk management team.
Procurement management	The length of the project may make it difficult to accurately determine procure-ment costs upfront.
Stakeholder management	Stakeholders must be in agreement with the alignment of the projects to strategic business objectives.

8.4 GENERIC STRATEGIC PLANNING

Figure 8-1 shows the framework by which many companies perform strategic planning. The box that seems to attract the least amount of attention is the last box, namely strategy implementation. This is where project management can provide the greatest value. Several years ago, a Fortune 500 company hired a team of consultants to help them with strategic planning. After the consultants left, the executives met and asked the critical question, "What did we learn from the consultants?" The executives realized that the consultants told them "what to do" but not "how to do it." Recognizing that implementation is what turns ideas into reality, the executives concluded that effective project management would be needed for strategic planning implementation.

Figure 8-2 shows the hierarchy of responsibilities. Once the high-level strategy is formulated, the strategy must be decomposed into manageable activities or projects, each with its own set of interim objectives. A business case may be required for each project and is usually prepared by the functional managers in coordination with the project managers. The wording of the business case has to be clearly understood by the project team. Executives may wish to sign off on each business case.

One of the characteristics of PM 2.0 is the ability to handle evolving requirements. Strategic planning projects have time horizons that may make a clear definition of the initial requirements impossible. Flexibility and the ability to handle continuous change will happen on many PM 2.0 projects.

Traditionally, the definition of success on a project was getting the work accomplished within time and cost and getting the final results accepted by the client. The definition of project success today, which makes it more appropriate for strategic planning activities, is the creation of sustainable business value acceptable to the clients and stakeholders. This definition induces the project manager to be more business oriented.

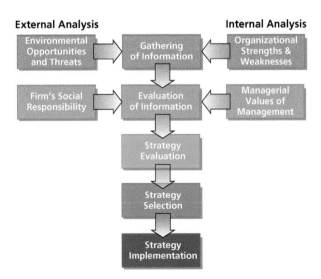

External Analysis

Environmental Opportunities and Threats → Gathering of Information ← Organizational Strengths & Weaknesses **Internal Analysis**

Firm's Social Responsibility → Evaluation of Information ← Managerial Values of Management

Strategy Evaluation

Strategy Selection

Strategy Implementation

FIGURE 8-1 **Traditional strategic planning activities**

FIGURE 8-2 **Hierarchy of responsibilities**

Before action plans can be formulated, the project manager must make sure that sufficient and qualified resources are available and project staffing can be accomplished in a timely manner. Executives usually look at the company's total head count, whereas project managers focus on the specific skills required, availability of resources whether full time or part time, whether outsourcing will be needed, and whether sufficient funding is available for the resources needed. Getting functional managers to make commitments for resources for the duration of the strategic project may be almost impossible.

Every company has its own approach to strategic planning and the categorization of projects. Previously discussed in Figure 4-2, we showed a simple classification system based upon the creation of business value. Each project selected must create sustainable business value for the company or else the project should not be part of the portfolio of projects. The classification of projects assists the project manager in determining the type of resources required. Typical classification categories include:

Internal Value: These projects are designed to improve the efficiency and effectiveness of the firm. The value obtained from these projects could be lowering costs, reducing waste, and shortening the time to market for new products. These projects can also be to improve the enterprise project management methodology, in which case people with process skills would be needed.

Financial Value: Companies need cash flow to survive. These projects could be to find better ways to market and sell the firm's products and services, in which case people with marketing and sales knowledge would be beneficial.

Customer-Related Value: The near-term value in these projects is that they improve customer relations. It is not uncommon for near-term projects to drain cash rather than generate cash. The long-term value comes from future contracts to support cash flow. Resources needed on these projects are generally people who know the customer or may have worked on projects for the customer previously.

Future Value: These projects are designed to create future value through new products and services. In most companies, the best technically oriented people are assigned to these projects based upon the subcategories. These projects may be heavily oriented around R&D. Typical subcategories might be radical breakthrough, next generation, addition to the family or add-ons, and enhancements. Future value projects may require project managers with technical skills as well as business skills and a good understanding of business risk management. Figure 8-3 shows the subcategories and the associated risk that must be continuously evaluated.

There are other factors that may be critical in selecting project resources, such as the payback period for the project. This is shown in Figure 8-4. The shorter the payback period, the more experienced the resources that the project manager needs. One of the responsibilities of today's PMOs is to support senior management in establishing the portfolio of projects. In this regard, the PMO will look at all of the projects together and determine the total resources needed as well as the appropriate skill levels. We must remember that the firm also needs qualified resources to support the ongoing business. This often creates a conflict for functional managers who must determine which resources can be released from ongoing work to support projects.

Product Changes	Radical Breakthrough	Next Generation	Addition to Family	Add-ons & Enhancements
Process Changes	New Process	Next Generation	Changes and Upgrades	Fine-Tuning & Incidentals
Risk Type				
• Market	• High	• High	• Medium	• Medium
• Technical	• High	• High	• Medium	• Medium
• Timing	• High	• High	• Medium	• Low
• Cost	• Low	• Medium	• Medium	• Low
• Price	• Medium	• Medium	• Low	• Low
• Quality	• Medium	• Medium	• Medium	• Low

FIGURE 8-3   Sample risk intensity for future-value projects

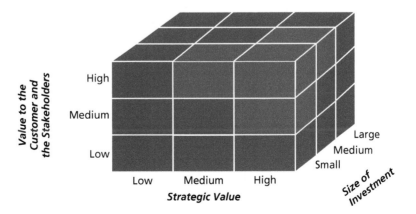

FIGURE 8-4   Typical resource analysis model

Since companies do not have an infinite supply of resources, the PMO will then make recommendations to senior management on how the project should be prioritized. Optimization techniques are available for the capacity-planning and prioritization efforts. But without a good business case for each project, we can easily end up with suboptimal solutions. We must also remember that it may not be possible for the project manager to hold onto all of the qualified resources for the duration of the project. Workers can get promoted off of your project or simply feel that they have better career path opportunities elsewhere.

8.5 BENEFITS OF PROJECT MANAGEMENT

Other than the benefits provided previously, perhaps the primary benefit of using project management, which makes it extremely attractive for strategic planning projects, is to provide executives and clients with a single point of contact for status reporting. Most of today's strategic planning projects are so complex that they cannot be managed effectively by one functional manager, who may have a conflict between his or her functional duties and project duties. These projects require the coordinated effort of several functional areas such as sales, marketing, engineering, and manufacturing. Without having a single point of contact for status reporting, the executives would need to do the coordination and integration themselves, and it is highly unlikely that they would have the time to do this in addition to their other duties. Likewise, functional managers do not have sufficient time to manage their functional areas and perform integration work on various projects. The need for project management is quite clear.

There are also many other benefits of using project management, some of which are shown in Table 8-2.

TABLE 8-2 Benefits of Using Project Management	
Attribute	**Benefit**
Efficiency	Allows an organization to take on more work in less time without any increase in cost or degradation of quality
Profitability	With all other things being equal, profitability should increase
Scope changes	Allows for better upfront planning, which should reduce the number of scope changes downstream and prevent unwanted changes from happening
Organizational stability	Focuses on effective teamwork, communication, cooperation, and trust rather than organizational restructuring
Quality	Quality and project management are married together; they both emphasize effective upfront planning
Risks	Allows for better identification and mitigation of risks
Problem solving	The project management processes allow for informed decision making and problem solving in a timely manner

TABLE 8-3 Additional Benefits for Strategic Planning Execution Projects

Attribute	Benefit
Alignment	Better alignment of projects to corporate strategic objectives
Underperformance	Earlier identification of underperforming investments
Capacity planning	Better analysis of corporate resource planning and availability of qualified resources
Prioritization	Combining capacity-planning efforts and project management allows for better prioritization of the portfolio of projects
Risk mitigation	Allows for better mitigation of business risks by using more "what if" scenarios
Time to market	Allows for quicker time to market
Decision making	More informed and timely decisions due to availability of essential information
Efficiency and effectiveness	Allows us to work on more projects without increasing head count
Better information flow	Elimination of duplication of efforts by managers who are unaware what others are doing
Selection of projects	Better analysis of what is and what is not a good idea

The benefits shown in Table 8-2 apply to just about all projects, including strategic planning execution projects, complex projects, and traditional projects. But there are some additional benefits that affect strategic planning execution projects more so than other types of projects. These are illustrated in Table 8-3.

8.6 DISPELLING MYTHS

When we look at Tables 8-2 and 8-3 and see all of the advantages, we must ask ourselves, "Why is there still resistance to the acceptance of project management, especially for strategic planning execution projects?" The answer is quite clear; there are still myths about the use of project management.

Myth 1: Project managers have strong technical knowledge but limited knowledge about the business. While it is true that historically project managers came out of technical disciplines and many even possessed master's degrees and doctorates in technical disciplines, today's project manager has more of an understanding of technology rather than a command of technology but an excellent knowledge of the business. Business knowledge is essential to bridge strategy and execution effectively. Project managers who are considered "global" project managers must have a good understanding of the client's business as well as their own firm's business. This is a necessity to compete in a global marketplace. These global project managers are also being trained in stakeholder relations management, politics, culture, and religion, since all of these topics can have an impact on the client's project.

We believe today that we are managing our firm's business as though it is a series of projects, where project managers are expected to make both project decisions and business decisions. Some companies are requiring their project managers to become certified in the company's business processes or to take coursework leading to certification as a business analyst.

Myth 2: Project managers should be assigned to a project after the project is approved and the business case is developed. Years ago, project managers were brought on board a project at the end of the initiation, rather than at the beginning. We believed that, because project managers had limited knowledge of the business, they could not contribute anything worthwhile during the initiation process. After the projects were selected, project managers would be brought on board and told to begin execution. Today, project managers are brought on board at the beginning of the project initiation and selection process and are expected to make a valuable contribution because of their understanding of the business.

Without effective project management practices, meaningful project metrics/KPIs, and accompanying organizational process assets that could be used for effective status reporting, it became difficult to cancel projects in a timely manner. Waiting too long, and once capital expenditures were incurred on projects, the costs associated with a canceled project could be significant. Without meaningful project performance metrics, it was often difficult to terminate a project in a timely manner and assign the resources to other projects with a greater opportunity for commercial success.

Bringing project managers on board early allows us to make better decisions in the selection of projects. The process by which this takes place is shown in Figure 8-5. The greatest contributions that the project manager can make are most likely with regard to impact analysis and risk assessment. The project manager may also be allowed to vote in the final go/no-go decision. By using the processes in Figure 8-5, we may need just

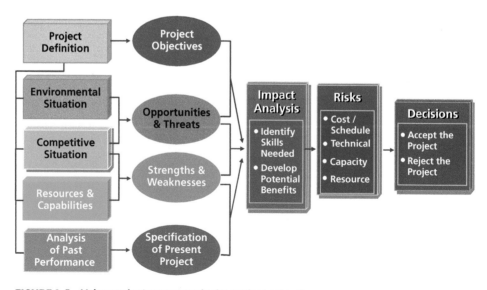

FIGURE 8-5   Using project managers in the project selection process

20 or fewer ideas to create a commercially successful product rather than the historical number of 60. Project managers may also participate in the activities in the first two columns of Figure 8-5, based upon their business knowledge. Although we are not now advocating that project managers need a master's degree in business administration, this may very well be an indication of educational requirements for the future.

Myth 3: If we implement project management, project managers will begin making decisions that should be made at the executive levels of management. Executives are ultimately responsible for strategic planning for a firm. They cannot and should not have someone else do it for them. However, in some cases strategic planning execution decisions may be made for the executives rather than by them. Executives have always been fearful of having to empower project managers with authority and responsibility with regard to project decision making. This myth alone has been a great impediment to the successful implementation of project management.

The problem was partially resolved with the creation of the position of the executive sponsor or project sponsor. Project managers were allowed to make technical decisions, but the project sponsors reserved the right to make any and all business-related decisions. This approach worked well for reasonably short duration projects. But for strategic planning execution projects, which can be 5–10 years in length, the number of decisions that must be made can be overwhelming. Therefore, to overcome this myth, it is beneficial to clearly define the empowerment of the project manager with regard to responsibilities and decision-making authority.

Myth 4: Project managers do not know how to use the organization process assets effectively for controlled measurement systems needed for informed decision making to take place. For the past five decades, the two primary metrics used by project managers were time and cost. This was because of the rule of inversion, which states that we often select the easiest metrics to measure and report, even though they may not provide us with a clear picture of the health of the project. Time and cost alone cannot predict the success of a project or whether the value will be there at the completion of the project. This is particularly true for strategic planning execution projects.

There are seminars in the marketplace today on measurement techniques. There are also textbooks on measurement techniques, which argue that anything can be measured if you simply understand the information at your disposal. The result has been the creation of additional metrics for project management. There is a belief that we should consider the following as core metrics for today's projects:

- Time
- Cost
- Resources
- Scope
- Quality
- Action items

These core metrics apply to all projects, but additional core metrics must be added based upon the size, nature, scope, and importance of the project. Because strategic planning implementation projects can be long in duration, significant changes can take place. As such, we must allow metrics to change over the course of the project. Establishing a set of core metrics that can be used on every project may be difficult.

8.7 WAYS THAT PROJECT MANAGEMENT HELPS STRATEGIC PLANNING

There are often special situations where project management can significantly benefit an organization. In a company that manufactures household appliances, each functional area was allowed to perform their own strategic planning. The problem occurred when functional units had to work together on the same project. In this company, new products were introduced at trade shows, and there were two trade shows each year. Missing a trade show product launch could easily result in lost revenue for six months until the next trade show.

The launching of new products was the highest priority in marketing's strategic plan. R&D, on the other hand, had more than 300 projects in the queue. On the R&D list of priorities, the new products that marketing needed for the trade shows were low on their list of priorities. Battles between marketing and R&D ensued continuously.

In another company, marketing was allowed to prioritize projects as part of their strategic planning activities. For each project, they also prioritized the attributes of the project/product that had a direct bearing on the way the product would be advertised and marketed. An example of this is shown in Figure 8-6. But when the project/product went into manufacturing, the manufacturing people often had a different set of priorities for the attributes, as shown in Figure 8-7 This time, battles over priorities ensued between marketing and manufacturing.

In both of the above-mentioned examples, the issues were resolved when project management personnel requested that the organization create a single priority list for all projects in the company. The result was that R&D, engineering, and manufacturing would meet once every three months and come to an agreement on the priorities of the projects. However, there were too many projects in the queue to prioritize each project. The decision was then made that only 20 projects at a time would be prioritized. This greatly benefited the project staffing process, because everyone was now working off of the same priority list.

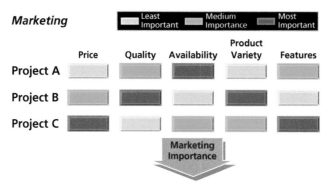

FIGURE 8-6 Marketing's attribute priority list

Manufacturing

	Least Important	Medium Importance	Most Important

	Cost Reduction	Hard Automation	Rate of Changes	Quality	Innovation/ Technology
Project A					
Project B					
Project C					

FIGURE 8-7 **Manufacturing's attribute priority list**

Another effective use of project management is gap analysis and gap closure. Gap analysis is used to strengthen your company's competitive position or to reduce the competitive position of your competitors by reducing gaps. Projects are established to take advantage of best practices and lessons learned on other projects, by which gaps can be compressed. The gaps can be:

- Speed with which new products are introduced (time to market)
- Competitiveness on cost
- Competitiveness on quality
- Introduction of new technology or product performance

Figure 8-8 shows the gap in new product development time between your firm and the industry average. The same curve can be used for competitiveness on cost. Figure 8-9 shows the gap between your firm and the industry leader. Therefore, strategic projects can be established to reduce your time to meet the industry average or compete with the industry leader.

Project management can also be used to manage strategic projects designed to improve quality. Figure 8-10 comes from an automotive supplier that was struggling to

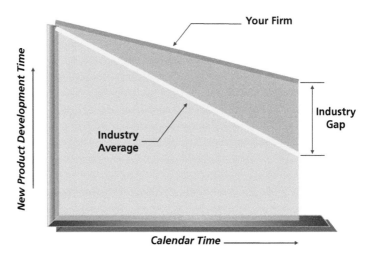

FIGURE 8-8 **Gap between your firm and the industry average**

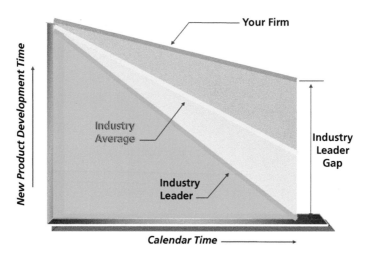

FIGURE 8-9 Gap between your firm and the industry leader

FIGURE 8-10 Identifying the quality gap

meet the quality demands imposed by their customers. A series of projects were then set up to improve quality.

Project management can also be used on projects designed to increase overall performance. In Figure 8-11, a company discovered that there was a gap in the overall performance of some of its products. They concluded that if the gap was not compressed in a reasonable amount of time, it could increase to a point where the company could be in serious trouble. A series of strategic projects was established to find ways to compress the gaps.

Executing strategic projects to compress performance gaps is difficult. These projects may require a great deal of innovation and brainstorming. Decisions may end up affecting the entire organization, and the final outcome may be a change in the corporate culture. And all of this is unknown when the project first begins.

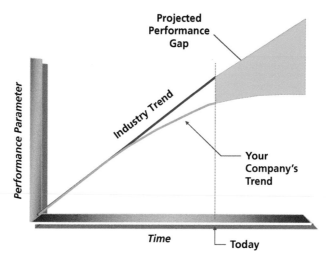

FIGURE 8-11   Identifying performance gaps

8.8 TRANSFORMATIONAL PROJECT MANAGEMENT LEADERSHIP

What type of leadership will project managers need for PM 2.0?

There have been numerous books written on effective project management leadership. Most books seem to favor situational leadership where the leadership style that the project manager selects is based upon the size and nature of the project, the importance of the deliverables, the skill level of the project team members, the project manager's previous experience working with these team members, and the risks associated with the project. Historically, project managers perceived themselves as being paid to produce deliverables rather than managing people. Team leadership was important to some degree as long as what was expected in the way of employee performance and behavior was consistent with the desires of the employee's functional manager that conducted the employee's performance review. In the past, project managers were expected to provide leadership in a manner that improves the employee's performance and skills and allows the employee to grow while working on project teams. All of this was based upon PM 1.0 using traditional project management practices.

Today, with PM 2.0, project managers are being asked to function as managers of organizational change on selected projects. Organizational change requires that people change. This mandates that project managers possess a set of skills that may be different than what was appropriate for managing projects. This approach is now being called transformational project management leadership.

Transformational leadership, according to Wikipedia, the free encyclopedia, enhances the motivation, morale, and performance of followers through a variety of mechanisms. These include connecting the follower's sense of identity and self to the project and the collective identity of the organization; being a role model for followers that inspires them and makes them interested; challenging followers to take greater ownership for their work, and understanding the strengths and weaknesses of followers, so the leader can align followers with tasks that enhance their performance.

Through the strength of their vision and personality, transformational leaders are able to inspire followers to change expectations, perceptions, and motivations to work toward common goals. Unlike other leadership approaches that may be based upon a "give and take" relationship, transformational leadership focuses on the leader's personality, traits, and ability to make a change through example, articulation of an energizing vision, and challenging goals. Transforming leaders are idealized in the sense that they are a moral exemplar of working toward the benefit of the team, organization, and/or community.

According to Wikipedia, there are four elements of transformational leadership:

Individualized Consideration: The degree to which the leader attends to each follower's needs, acts as a mentor or coach to the follower, and listens to the follower's concerns and needs. The leader gives empathy and support, keeps communication open, and places challenges before the followers. This also encompasses the need for respect and celebrates the individual contribution that each follower can make to the team. The followers have a will and aspirations for self-development and have intrinsic motivation for their tasks.

Intellectual Stimulation: Such leaders encourage their followers to be innovative and creative. They encourage new ideas from their followers and never criticize them publicly for the mistakes committed by them. The leaders focus on the "what" in problems and do not focus on the blaming part of it. They have no hesitation in discarding an old practice set by them if it is found ineffective.

Inspirational Motivation: The degree to which the leader articulates a vision that is appealing and inspiring to followers. Leaders with inspirational motivation challenge followers to leave their comfort zones, communicate optimism about future goals, and provide meaning for the task at hand. Followers need to have a strong sense of purpose if they are to be motivated to act. Purpose and meaning provide the energy that drives a group forward. The visionary aspects of leadership are supported by communication skills that make the vision understandable, precise, powerful, and engaging. The followers are willing to invest more effort in their tasks, and they are encouraged and optimistic about the future and believe in their abilities.

Idealized Influence: The degree to which the leader acts as a role model for their followers. Transformational leaders must embody the values that the followers should be learning and mimicking back to others. If the leader gives respect and encourages others to be better, those influenced will then go to others and repeat the positive behavior, passing on the leadership qualities for other followers to learn. This will earn the leader more respect and admiration from the followers, putting them at a higher level of influence and importance. The foundation of transformational leadership is the promotion of consistent vision, mission, and a set of values to the members. Their vision is so compelling that they know what they want from every interaction. Transformational leaders guide followers by providing them with a sense of meaning and challenge. They work enthusiastically and optimistically to foster the spirit of teamwork and commitment.

Each element is connected because there is a basis of respect, encouragement, and influence that is involved in transformational leadership. The personality of the leader has to be genuine because any chance of inconsistency for the followers and all trust is gone, and the leader has failed.

There is a valid argument that all project managers are managers of change and therefore should adopt a transformational project management leadership style. While this argument has merit, there are obstacles on some projects where this leadership style may not work.

- In the above four elements of transformational leadership, the word "follower" was used. In a project management environment, the workers are referred to as associates rather than followers or subordinates and may even be at a higher pay grade than the project manager. Some of the followers may even be executives working on the project.
- Transformational leadership may require that the project manager have some degree of authority to enforce any promises made to the workers. PM 1.0 is often seen as leadership without authority.
- Project managers may have little or no involvement in employee wage and salary administration and therefore cannot reward (or punish) employees for the way they respond to transformational leadership efforts.
- Team members may be assigned to the project team only a few hours a week and the project manager may have a difficult time using a transformational leadership style. It may be very difficult to get workers to leave their comfort and work differently when the project manager interfaces with the workers only a few hours a week at most.
- Team members may be assigned to multiple projects and each project manager may adopt a different leadership style.

However, there are specific situations where transformational leadership must be used and employees must be removed from their previous comfort zones. As an example, not all projects come to an end once the deliverables are created. Consider a multinational company that establishes an IT project to create a new, high-security companywide e-mail system. Once the software is developed, the project is ready to "go live." Historically, the person acting as the project manager to develop the software moves on to another project at go live, and the responsibility for implementation goes to the functional managers or someone else. Today, companies are asking the project manager to remain on board the project and act as the change agent for full, corporatewide implementation of the change-over to the new system. In these situations, the project manager must adopt a transformational leadership style.

In a project management context, change management usually refers to managing the project management processes associated with the approval and implementation of project scope changes. But in the above go-live example, the project manager is now responsible for organizational change management for ensuring that the changes are successfully implemented to achieve the expected lasting benefits. So as not to confuse the two change management terms, organizational change management mandates the use of transformational leadership where the goal is to transform the organization from one state to another.

Transformational project management is heavily focused upon the people side of the change and is a method for managing the resistance to the change, whether the change is in processes, technology, acquisitions, targets, or organizational restructuring. People need to understand the change and buy into it. Imposing change upon people is an invitation for prolonged resistance, especially if people see their job threatened. Transformational projects can remove people from their comfort zones.

Before discussing the skills needed to be a transformational project manager, it is important to understand the transformational environment. Factors that can influence the transformational environment include:

- Having to work in a multinational setting and interfacing with different cultures
- Recognizing alignment to broader business objectives
- Creating an environment conducive to change
- Overcoming employee resistance to change
- Understanding worker sensitivity to the changes
- Winning employee support for the changes
- Assessing what help you need to achieve the changes
- Making people believe that they are not being manipulated
- Making sure that people can cope with the changes
- Diffusing emotional feelings
- Holding face-to-face meetings to verify that people understand the need for the change
- Using workshops when necessary
- Providing training for the changes
- Managing bad news effectively
- Managing unprecedented scrutiny by stakeholders
- Managing scrutiny by the media

The skills needed to manage transformational projects can be significantly different than traditional project management skills as indicated in Table 8-4.

TABLE 8-4 Differences between Traditional and Transformational Project Management Leadership

Traits	Differences Related to Transformational Projects
Authority	From leadership without authority to significant authority
Power	From legitimate power to judicious use of power
Decision making	From some decision making to having authority for significant decision making
Types of decisions	From project-only decisions to project and business decisions
Willingness to delegate	The length and size of the project will force the project managers to delegate more authority and decision making than they normally would
Loyalty	From project loyalty to corporate vision and business loyalty
Social skills	Strong social skills are needed to overcome the resistance to change

TABLE 8-4 *(Continued)*	
Traits	**Differences Related to Transformational Projects**
Motivation	Learning how to motivate workers without using financial rewards and power
Communication skills	Communication across the entire organization rather than with a selected few
Status reporting	Recognizing that the status of strategic projects cannot be made from time and cost alone
Perspective/outlook	Having a much wider outlook, especially from a business perspective
Vision	Must have the same long-term vision as the executives and promote the vision throughout the transformational effort
Compassion	Must have a much stronger compassion for the workers since they may be required to leave their comfort zones
Self-control	Must not overreact to bad news or disturbances
Brainstorming and problem solving	Must have very strong brainstorming and problem-solving skills
Change management	Going from project to corporatewide change management
Change management impact	Going from project to organizational change management effects

8.9 PROJECT MANAGER'S ROLE IN PORTFOLIO MANAGEMENT

Project selection decision makers frequently have much less information to evaluate candidate projects than they would wish. Uncertainties often surround the success likelihood of a project, the ultimate market value of the project, the skill level of the resources needed, the governance that may be necessary, and the project's total cost to completion. The lack of an adequate information base often leads to another difficulty: the lack of a systematic approach to project selection and evaluation.

Consensus criteria and methods for assessing each candidate project against these criteria are essential for rational decision making. Though most organizations have established organizational goals and objectives, these are usually not detailed enough to be used as criteria for project selection decision making. However, they are an essential starting point.

There are several benefits of having project managers participate in the project selection process. Project managers can identify:

- Potential risks in the execution of the project
- The skill level of the resources needed
- Whether the targeted cost and schedule seem to be realistic
- What level of priority might be needed

Project managers can also participate in the financial analysis activities with the feasibility study and benefit-to-cost analysis. The purpose of the feasibility study is to validate that the idea or project meets feasibility of cost, technological, safety, marketability, and ease of execution requirements. It is possible for the company to use outside consultants or subject matter experts (SMEs) to assist in both feasibility studies and benefit-to-cost analyses if the project manager may not have sufficient business or technical knowledge to contribute at this time. In addition, the project manager will get a better understanding of the constraints, assumptions, and enterprise environmental factors. This could eliminate downstream guesswork if the business case is not prepared adequately.

Project selection and evaluation decisions are often confounded by several behavioral and organizational factors. Departmental loyalties, conflicts in desires, differences in perspectives, and an unwillingness to openly share information can stymie the project election and evaluation process. Such project evaluation data and information are necessarily subjective in nature. Thus, the willingness of the parties to openly share and put trust in each other's opinions becomes an important factor. The risk-taking climate or culture of an organization can also have a decisive bearing on the project selection process. If the climate is risk averse, then high-risk projects may never surface. Attitudes within the organization toward ideas and the volume of ideas being generated will influence the quality of the projects selected. In general, the greater the number of creative ideas generated, the greater the chances of selecting high-quality projects.

There must be careful documentation of all known constraints and assumptions that were made in developing the costs and the benefits. Unrealistic or unrecognized assumptions are often the cause of unrealistic benefits. The go or no-go decision to continue with a project could very well rest upon the validity of the assumptions.

Many organizations make the fatal mistake of taking on too many projects without regard for the limited availability of resources. As a result, the highly skilled labor is assigned to more than one project, creating schedule slippages, lower productivity, less than anticipated profits, and never-ending project conflicts. This is also where the project manager can provide meaningful input.

The selection and prioritization of projects must be made based upon the availability of qualified resources. Planning models are available to help with the strategic timing of resources. These models are often referred to as aggregate planning models. Aggregate planning models allow an organization to identify the overcommitment of resources. This could mean that high-priority projects may need to be shifted in time or possibly be eliminated from the queue because of the unavailability of qualified resources. It is a pity that companies also waste time considering projects for which they know that the organization lacks the appropriate talent.

8.10 VALUE MANAGEMENT AND BENEFITS REALIZATION

Organizations in both the public and private sectors have been struggling with the creation of a portfolio of projects that would provide sustainable business value. All too often, companies would add all project requests to the queue for delivery without proper evaluation and with little regard if the project were aligned with business objectives or provided benefits and value upon successful completion. Projects were often submitted without any accompanying business cases. Many projects had accompanying business cases that were based upon highly exaggerated expectations and unrealistic benefits. Other projects were created because of the whims of management and the order in which the projects were completed was based upon the rank or title of the requestor. Simply because an executive says "Get it done" does not mean it will happen. The result was often project failure and a waste of precious resources. In some highly visible and well-publicized cases, business value was eroded or destroyed rather than created.

With the growth of metrics and KPIs, especially value-reflective metrics as discussed in earlier chapters, it is highly beneficial to have the project managers participate in portfolio management activities. The project manager will have a better understanding of the value that is expected and the value attributes that are important to the executives. The project manager can also discuss with the executives how they wish to have the value-reflective metrics displayed. These discussions will mandate involvement by the PMO during the portfolio selection process. However, there are other metrics as well that should be discussed, as will be described in this section.

Understanding the Terminology

Before continuing on, it is important to understand the terminology.

A *benefit* is something that is considered to be important or advantageous to specific individuals or a group of individuals. Benefits, whether they are strategic or nonstrategic, are normally aligned to the organizational objectives of the sponsoring organization that will eventually receive the benefits. The benefits appear through the deliverables or outputs that are created by the project. It is the responsibility of the project manager to create the deliverables.

Benefits are identified in the project's business case. Some benefits are tangible and can be quantified. But it is more likely that the benefits are intangible at this stage in the project and difficult to quantify.

Benefits realization management is a collection of processes, principles, and deliverables to effectively manage the organization's investments. Project management focuses on maintaining the established baselines whereas benefits realization management analyzes the relationship that the project has to the business objectives by monitoring for potential waste, acceptable levels of resources, risk, cost, quality, and time as it relates to the benefits. It is entirely possible that the benefits can change over the life of the project to a point where the outcome of the project provides detrimental results.[1]

Project *value* is what the benefits are worth to someone. Project or business value can be quantified whereas benefits are usually explained qualitatively. On some projects, the value of the benefits of the project cannot be quantified until several months after the project has been completed. As an example, a government agency enlarges a road to hopefully reduce traffic congestion. The value of the project may not be known until several months after the construction project has been completed and traffic flow measurements have been made.

Benefits realization and business value do not come from simply having talented resources or superior capabilities. Rather, they come from how the organization uses the

---

1 For additional information, see C. Letavec, *Strategic Benefits Realization*, J. Ross Publishers, Plantation, FL, 2004, and T. Melton, P. Iles-Smith, and J. Yates, *Project Benefits Management; Linking Projects to the Business*, Butterworth-Heinmann, an imprint of Elsevier Publishers, Oxford, UK, 2008.

resources. Sometimes, even projects with well-thought-out plans and superior talent do not end up creating business value and can even destroy existing value.

Life-Cycle Phases

Typical project life-cycle phases begin once the project is approved and end after the deliverables have been created. However, when value management and benefits realization become important, there are additional life-cycle phases that must be included as shown in Figure 8-12. Figure 8-12 is more representative of an investment life cycle than a project life cycle. The project life cycle is contained within the investment life cycle. If value is to be created, then the benefits must be managed over the complete investment life cycle. More than six life-cycle phases could have been identified in the investment life cycle, but only these six will be considered here for simplicity.

The *idea generation phase* is where the idea for the project originates. The idea can originate in the client's organization, within the senior level or lower levels of management in the parent company or the client, or within the organization funding the project. The output of the idea generation phase is usually the creation of a business case that may include:

- Opportunities such as improved efficiency, effectiveness, waste reduction, cost savings, new business, etc.
- Benefits defined in both business and financial terms
- A benefit realization plan if necessary
- Project costs
- Recommended metrics for tracking the benefits
- Risks
- Resource requirements
- Schedules and milestones
- Complexity
- Assumptions and constraints
- Technology requirements; new or existing technology
- New business opportunities
- Exit strategies if the project must be terminated

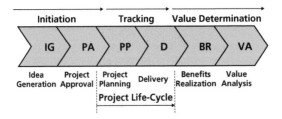

FIGURE 8-12 Investment life cycle

Although the idea originator may have a clear picture of the ultimate value of the project, the business case is defined in terms of expected benefits rather than value. Value is determined near the end of the project based upon the benefits that are actually achieved and quantified. The benefits actually achieved may be significantly different from the expected benefits defined at project initiation.

Not all projects require the creation of a business case. Examples might include projects that are mandatory for regulatory agency compliance or simply to allow the business to continue.

Once the business case is prepared, a request is sent to the PMO for project approval. Companies today are establishing a portfolio management PMO to control the *project approval phase* and to monitor the performance of the portfolio of projects during delivery.

The PMO must make decisions for what is in the best interest of the entire company. A project that is considered as extremely important to one business unit may be a low priority when compared to all of the other corporate projects in the queue. The PMO must maximize the benefits through proper balancing of critical resources and proper prioritization of projects.

The portfolio PMO must address three critical questions as shown in Table 8-5.

TABLE 8-5 Typical Role for a Portfolio PMO

Critical Questions	Areas of Consideration	Portfolio Tools and Processes
Are we doing the right things?	• Alignment to the strategic goals such as shareholder value or customer satisfaction • Evaluation of internal strengths and weaknesses • Evaluation of available and qualified resources	• Templates to evaluate rigor of the business case • Strategic fit analysis and linkage to strategic objectives • Matrix showing the relationships between projects • Resources skills matrices • Capacity-planning templates • Prioritization templates
Are we doing enough of the right things?	• Comparison to strategic goals and objectives • Ability to meet all of the customers' expectations	• Overall benefits tracking • Accurate reporting using the project management information system
Are we doing the right things right?	• Ability to meet expectations • Ability to achieve benefits • Ability to manage technology	• Benefit realization plans • Formalized detailed project plans • Establishing tracking metrics and KPIs • Risk analysis • Issues management • Resource tracking • Benefits tracking

The activities identified with the third question in Table 8-5 are usually part of the PMO's responsibility for monitoring performance once the project is approved.

The third life-cycle phase is the *project planning phase*. This phase includes preliminary planning, detailed planning, and benefits realization planning. Although the business case may include assumptions and constraints, there may be additional assumptions and constraints provided by the PMO related to overall business objectives and the impact that enterprise environmental factors may have on the project. The benefits realization plan that may have been created as part of the business case may undergo significant changes in this phase based upon the project manager's ability to predict risks in project execution.

The benefits realization plan is not the same as the project plan but must be integrated with the project plan. The benefits realization plan may undergo continuous change as the project progresses based upon changing business conditions. Items that may be unique to the benefits realization plan include:

- A description of the benefits
- Identification of each benefit as tangible or intangible
- Identification of the recipient of each benefit
- How the benefits will be realized
- How the benefits will be measured
- The realization date for each benefit
- The handover activities to another group that may be responsible for converting the project's deliverables into benefits realization

The fourth life-cycle phase is the *delivery phase*. This phase is most commonly based upon the ***PMBOK® Guide*** standards or any other project management standards. Traditional project management methodologies are used. In this phase, the project manager works closely with the PMO and the steering/governance committee to maximize the realization of the benefits.

Performance reporting must be made available to the portfolio PMO as well as to the appropriate stakeholders. If the alignment of the project with business objectives has changed during delivery, the PMO may recommend that the project be redirected or even canceled such that the resources will then be assigned to other projects that can provide a maximization of benefits.

There are numerous factors which may lead to project cancellation. Some of these, which have already been included in Section 8.1, are:

- During project delivery, neglecting to recognize changes in the enterprise environmental factors and how they influence the benefits expected as well as senior management's vision of the future
- Unfavorable changes in the assumptions and constraints
- Exaggerated or unrealistic benefits and value established during project approval
- Poorly defined or ill-defined benefits
- Poorly documented business case resulting in the approval of the wrong project
- Failing to get executive and stakeholder buy-in right from the start
- Poor executive governance resulting in lack of support and poor decision making

- Constantly changing the membership of the governance team resulting in a constant change of project direction
- Overestimating resource competencies needed for project delivery
- Poor capacity-planning efforts resulting in an understaffed project or a project staffed with resources lacking the necessary skills
- Functional managers refusing to commit the proper resources for the duration of the project
- Failing to get employee commitment to the project and the expected benefits
- Failing to explain the project well to the project delivery team
- Failing to understand the magnitude of the organizational change needed for the benefits and value to be achieved
- Unable to manage change effectively
- Failing to consider the impact of changes in technology during the delivery of the project
- Poor estimating of time and cost
- Having an execution team that is unable to work with ill-defined or constantly changing requirement
- Poor integration of the project across the entire organization
- Inadequate communications throughout the organization

The last two life-cycle phases in Figure 8-12 are the *benefits realization phase* and the *value analysis phase*. The benefits realization plan, regardless in which life-cycle phase it is prepared, must identify the metrics that will be used to track the benefits and accompanying value. Benefits and value metrics are the weak links in benefits realization planning. Much has been written on the components of the plan but very little appears on the metrics to be used.[2]

Table 8-6 identifies four broad categories that can be used for the identification of benefits. These are the same four categories identified in Figure 4-3. There are numerous benefits and metrics that can be used for each category. Only a few appear here as examples.

Metrics serve as early warning signs of possible problems. Some examples might be:

- Metrics on the number of scope changes identify the possibility of a schedule slippage and cost overrun.
- Metrics on the number of people removed to put out fires elsewhere also indicate the possibility of a schedule slippage and cost overrun.
- Metrics on excessive overtime could indicate serious issues.
- Metrics on missed deadlines indicate that the time-to-market may slip and opportunities may be lost.

The project tracking metrics identified in Table 8-6 are designed to track individual projects in each of the categories. However, there are specific metrics that can be used to measure the effectiveness of a PMO. Table 8-7 shows the metrics that can be used

---

[2] For information on creating and reporting value metrics, see H. Kerzner, *Project Management Metrics, KPIs and Dashboards*, 2nd ed., Wiley, Hoboken, NJ, 2013, Chapter 5.

TABLE 8-6 Typical Categories of Benefits and Value

Category	Benefits	Project Tracking Metrics
Internal benefits	• Adherence to constraints • Repetitive delivery • Control of scope changes • Control of action items • Reduction in waste • Efficiency	• Time • Cost • Scope • Quality • Number of scope changes • Duration of open action items • Number of resources • Amount of waste • Efficiency
Financial benefits	• Improvements in ROI, NPV, IRR, and payback period • Cash flow • Improvements in operating margins	• Financial metrics • ROI calculators • Operating margins
Future benefits	• Reducing time to market • Image/reputation • Technical superiority • Creation of new technology or products	• Time • Surveys on image and reputation • Number of new products • Number of patents • Number of retained customers • Number of new customers
Customer-related benefits	• Customer loyalty • Number of customers allowing you to use their name as a reference • Improvements in customer delivery • Customer satisfaction ratings	• Loyalty/customer satisfaction surveys • Time to market • Quality

to measure the overall value of project management, a traditional PMO, and a portfolio PMO. The metrics listed under project management and many of the metrics under the traditional PMO are considered as microlevel metrics focusing on tactical objectives. The metrics listed under the portfolio PMO are macrolevel metrics. Both the traditional and portfolio PMOs are generally considered as overhead and subject to possible downsizing unless the PMOs can show through metrics how the organization benefits by their existence.

It is important to understand that some of the micro metrics we use for tracking benefits may have a different meaning for the customer. As an example, let us assume that you are managing a project for an external client. The deliverable is a component that your customer will use in a product he or she is selling to their customers (i.e., your

TABLE 8-7 Comparison of Project Management and PMO Metrics		
Project Management (Micro Metrics)	**Traditional PMO (Macro Metrics)**	**Portfolio PMO (Macro Metrics)**
• Adherence to schedule baselines • Adherence to cost baselines • Adherence to scope baselines • Adherence to quality requirements • Effective utilization of resources • Customer satisfaction levels • Project performance • Total number of deliverables produced	• Growth in customer satisfaction • Number of projects at risk • Conformance to the methodology • Ways to reduce the number of scope changes • Growth in the yearly throughput of work • Validation of timing and funding • Ability to reduce project closure rates	• Business portfolio profitability or ROI • Portfolio health • Percentage of successful portfolio projects • Percentage of projects that failed to deliver • Percentage of projects that were stopped • Percentage of benefits realized • Portfolio value achieved • Portfolio selection and mix of projects • Resource availability • Capacity available for the portfolio • Utilization of people for portfolio projects • Hours per portfolio project • Staff shortage • Percent with strategic alignment • Business performance enhancements • Business opportunities • Business outcomes • Portfolio budget versus actual • Portfolio deadline versus actual • ROI met and forecasted • Portfolio risk levels • Percent of projects to run the business • Percent of projects to grow the business • Percent of projects requiring innovation • Percent of projects that are long, medium, and short term

TABLE 8-8 Interpretation of Metrics

Benefit Metric	Project Manager's Interpretation	Customer's Interpretation	Consumer's Interpretation
Time	Project duration	Time to market	Delivery date
Cost	Project cost	Selling price	Purchasing price
Quality	Performance	Functionality	Usability
Technology and scope	Meeting specifications	Strategic alignment	Safe buy and reliable
Satisfaction	Customer satisfaction	Consumer satisfaction	Esteem in ownership
Risks	No future business from this client	Loss of profits and market share	Need for support and risk of obsolescence

customer's customers or consumers). Table 8-8 shows how each of the metrics may be interpreted. It is important to realize that benefits and value are like beauty; they are in the eyes of the beholder. Customers and contractors can have a different perception of the meaning of benefits and value.

Understanding Value

Value is what the benefits are worth either at the end of the delivery phase or sometime in the future. Even though the benefits may be on track for achievement, the value can change. Consider the following examples:

A company approves the development of a customized software package with the expected benefit of reducing order entry processing by 50%, which would be a savings of approximately $1 million annually. The cost of developing the package is estimated at $3 million. As the project delivery process begins, the company realizes that the cost of developing the software will be closer to $5 million rather than $3 million. Even though the benefit of a cost savings still exists, the value has diminished and the payback period is now five years rather than three years. The company might now consider canceling the project and assigning resources elsewhere.

Lesson learned: Even though the benefits may not have changed, there can be an unfavorable change in the expected value.

A company has a contract to manufacture a component for a customer. The benefit is an expected $100,000 in profit on this contract. The resources could be assigned to a second project where the expected profit is greater. However, the customer for the first project is expecting to make $100 million in profits from the sale of their products that include your component. If the customer is happy with your component, you could very well receive a significant number of follow-on contracts such that the long-term value of the first project significantly supports long-term strategic objectives.

Lesson learned: Even though the benefits have not changed, there can be a significant favorable change in the expected value.

A company is working on an internal project with a well-defined benefits realization plan. During the delivery of the project, the costs are escalating and the schedule has been extended in order to provide added value to the internal customer. The customer authorized the cost overrun and schedule slippage. The benefits realization plan is updated.

Lesson learned: Benefits and value can change as the delivery of the project takes place over the time horizon.

8.11 BENEFITS REALIZATION METRICS

In Table 8-7 we listed several metrics that can be used by the portfolio PMO for displaying information on benefits and value. It should be obvious that there will be a cost incurred with capturing the information. Figure 8-13 shows that the cost of obtaining the information must be balanced against the value of having this information. Information below the "parity line" will almost always be worth the effort. Information above the parity line should be considered on an individual basis.

In order to use the macro metrics in Table 8-7, information must be obtained from each individual project. This is where the project manager and the portfolio PMO interface. Figure 8-14 shows a typical scoring model to determine the value for one project. Figure 8-15 shows how the points can be assigned to perform the evaluation. The points can and will be different for each project. In Figure 8-15, the project received only 80 points out of 100, and business value achieved was only 57 points out of a possible 70 points.

If we combine all of the scores for each project, we can display the yearly results as shown in Figure 8-16. In this case, the portfolio PMO is seeing continuous improvement each year in project benefit realization. Charts like Figure 8-16 are a necessity to show that the PMO is adding value to the overall business. PMOs often add significant value but do not measure it or report it. Unless it is reported, there is a risk that executive decision makers may downsize or eliminate the PMO considering it as overhead. Every PMO will have its own distinct way of showing its contribution to the business.

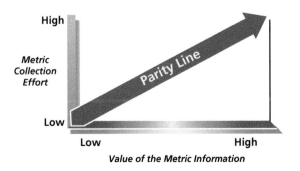

FIGURE 8-13   Evaluating the cost of metric information

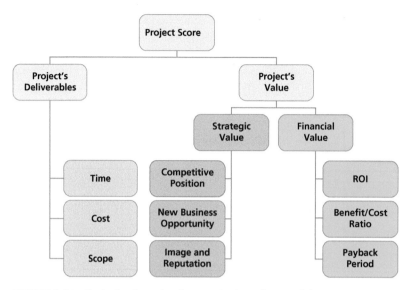

FIGURE 8-14 **Typical categories for a project scoring model**

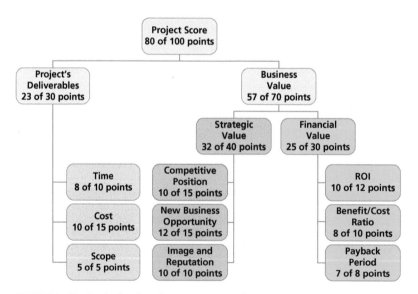

FIGURE 8-15 **Typical points for a project scoring model**

Another important metric that executives want to see is the alignment of the projects' objectives to the strategic business objectives. This can be done using an objectives matching matrix as seen in Figure 8-17. Points can be assigned to each cell and the scores can be totaled for each row and each column.

Obviously, all of the macro portfolio PMO metrics can be displayed. But care must be taken to determine the amount of time and money expended to do this.

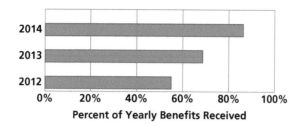

FIGURE 8-16   Yearly benefits realized

Strategic Objectives:	Project 1	Project 2	Project 3	Project 4	Project 5	Project 6	Project 7	Project 8	Scores
				Projects					
Technical Superiority	2		1			2		1	6
Reduced Operating Costs				2	2				4
Reduced Time-to-Market	1		1	2	1	1		2	8
Increase Business Profits			2	1	1	1		2	7
Add Manufaturing Capacity	1		2	2		1		1	7
Column Scores	4	0	6	7	4	5	0	6	

	= No Contribution
1	= Supports Objective
2	= Fulfills Objective

FIGURE 8-17   Objectives match matrix

8.12 PORTFOLIO MANAGEMENT GOVERNANCE

Every portfolio PMO has its own structure, responsibilities, and governance. One size does not fit all. No two PMOs are identical. Yet even with the differences, there are five fundamental questions that need to be answered for each portfolio PMO:

- When will the projects be completed?
- What will be the completion costs?
- How are the projects aligned to strategic objectives?
- What mix of projects maximizes the value to the business?
- What impact will the portfolio have on shareholder value?

With PM 2.0, membership of the portfolio governance committee must clearly understand their roles and responsibilities. The project managers must also understand how to interface with the oversight group because many of the people performing oversight may never have worked as a project manager. Perhaps the most important role is to evaluate and compare the project execution risks against the desired business value. Input from the project managers will be essential. Figure 8-18 shows the interfacing

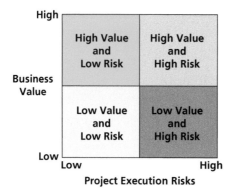

FIGURE 8-18   Mapping business value risks

between the project manager and the governance committee for mapping risks against business value.

Having a portfolio where all of the projects are high risk is dangerous. Yet some companies will see this as the pot of gold at the end of the rainbow and accept the risks.

There are several possible corrective actions that the governance committee can take to lower the portfolio's risks or to manage poorly performing projects. These include:

- Re-baselining the portfolio
- Terminating or removing weak investments
- Recommending scope changes to some of the existing projects
- Cutting costs
- Accelerating some schedules
- Consolidating some projects
- Changing project personnel

One of the most important points, and one often not recognized, is that the governance committee establishes the culture for the portfolio management process. The committee must:

- Eliminate pet projects, even their own pet projects, from being part of the portfolio
- Avoid creating a "big brother is watching you" environment that may antagonize the portfolio team
- Make sure that the correct level of detail appears in the information they require
- Make sure that the information they receive is useful
- Make sure that the information is used for decision making
- Make sure they understand the cost of collecting the information

Because of the importance of benefits and value, today's PM 2.0 project managers are more business managers than the pure project managers of the past. Today's project managers are expected to make business decisions as well as project-based decisions. Project managers seem to know more about the business than their predecessors.

With the growth in measurement techniques, companies will begin creating metrics to measure benefits and value. While many of these measurement techniques are still in the infancy stages, the growth rate is expected to be rapid.

DISCUSSION QUESTIONS

The discussion questions are for classroom use to stimulate group thinking about PM 2.0. There are no right or wrong answers to most of the questions.

1. What role does project management play in strategic planning?
2. What type of strategic planning projects would be ideal for a project manager to manage?
3. Can a project that supports a strategic plan be successful and yet the strategy fails?
4. What are the differences between traditional project management leadership and transformational project management leadership?
5. What are the differences if any between value management and benefits realization?
6. Who is responsible for creating the benefits realization plan?
7. Who is responsible for approving the benefits realization plan?
8. Can the benefits realization plan change over the project life cycle?
9. Is it a good or bad idea for project managers to participate in the establishment of a portfolio of projects?
10. Should project managers ever be allowed to serve on the governance committee of a project while they are acting as the project manager on another project?

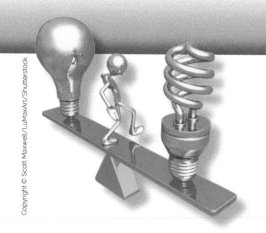

CHAPTER 9
R&D PROJECT MANAGEMENT

9.0 INTRODUCTION

For years, R&D project management performed in a PM 1.0 environment had been regarded as some type of mystique. The relationship between R&D, strategic planning and project management was not very well known. Companies even went so far as to develop enterprise project management methodologies that did not include R&D. One of the reasons for this was the belief that, if you can develop a schedule and budget for an R&D project, you do not have R&D. R&D is considered an environment where you may not know where you are going, you may not know the cost, and you may not know if you will ever get there. In most companies, the percentage of R&D projects that end up developing commercially successful products is quite low. Even though these same risks appear in PM 2.0, R&D projects will create the future for many companies. In PM 2.0, we must carefully select the projects to be worked on rather than simply adding all projects to the queue.

Very few people in the company seemed to understand the process by which R&D projects were selected and prioritized. Some people believed that Figure 9-1 was a representation of R&D project selection. Today, all of these beliefs are slowly going by the wayside thanks to the establishment of a PMO. The PMO is responsible for participating in the portfolio selection of projects, and this includes involvement in R&D-type projects as well. The PMO brings to the table a more structured process for the portfolio selection of projects. However, not all R&D projects require PMO involvement.

The need to work on more R&D projects is quite apparent. In a PM 2.0 environment, the need for effective R&D is expected to increase significantly. Without effective R&D, companies can become followers rather than industry leaders. And this can happen overnight.

PM 2.0 practices cannot guarantee that you will select the right R&D projects, nor can it guarantee that the projects selected will be successful. However, PM 2.0 practices can certainly improve your chances for success as well as identify early on those projects that should be canceled or redirected.

FIGURE 9-1   R&D selection process

9.1 ROLE OF R&D IN STRATEGIC PLANNING

Before looking at the various types of R&D projects, we must first understand the relationship between R&D and strategic planning. Strategic planning and R&D are similar in that both deal with the future profits and growth of the organization. Without a continuous stream of new products, the company's strategic planning options may be limited. Today, advances in technology, growing competitive pressures, and the necessity to compress innovative products is increasing while the life cycle of existing products appears to be decreasing at an alarming rate. Yet, at the same time, executives may keep research groups in a vacuum and fail to take advantage of the potential profit contribution resulting from effective R&D project management. The PMO appears to be making it easier to integrate R&D into the ongoing business while at the same time provide valuable information during the portfolio management of projects.

There are three primary reasons why corporations conduct R&D:

- To produce new products or services for profitable growth
- To produce profitable improvements to existing products and services
- To produce scientific knowledge that can assist in identifying new opportunities or in "fighting fires"

Each of these can require a different form of project management, but they all require R&D project management of some sort.

Successful R&D projects are targeted toward the accomplishment of specific objectives, but targeting requires a good information system and this, unfortunately, is the weakest link in most companies. Information systems are needed for optimum R&D targeting efforts, and this includes assessing customer and market needs, economic evaluation, and project selection. This is where the PMO may provide valuable assistance, but other groups, such as marketing, must also be involved.

Historically, with PM 1.0, most R&D project managers had technical backgrounds and may have had limited knowledge about the business. The result was often the creation of expensive products that the marketplace did not want or could not afford. Technology was created for the sake of technology and without fully understanding the needs of the business. With PM 2.0, the project managers are expected to make business as well as technical decisions, and this keeps decision making aligned with strategic business objectives. The role of the PMO is still critical to monitor the continuous alignment.

Assessing customer and market needs involves opportunity-seeking and commercial intelligence functions. Most companies delegate these responsibilities to the marketing group, and this may result in a detrimental effort because marketing groups appear to be overwhelmed with today's products and near-term profitability. They simply do not have the time or resources to adequately analyze other R&D activities that may have long-term implications. Also, marketing groups may not have technically trained personnel who can communicate effectively with the R&D groups of the customers and suppliers. Another problem is that marketing may have a "wish list" of projects they would like to see accomplished but resource limitations can exist. The PMO often has the responsibility to perform capacity-planning activities, thus providing valuable information on how much additional work can be undertaken without overburdening the existing labor force.

The success and possibly direction of a corporate strategic plan may be dependent solely upon that point in time where marketing tests the new products or features in the marketplace. Quite often, senior management will attempt to pressure the project managers to take risks and shorten the R&D time in order to increase earnings at a faster rate. The problem here is that the legality of product liability may be overlooked in the executives' haste to produce results due to the stockholder pressure. In time of trouble, executives cut R&D funding in the mistaken belief that cost reductions will occur. In either event, the long-term impact on the organization may prove unhealthy. What is unfortunate is that R&D project managers are often not provided with all of the market research data during project initiation, and this sometimes causes the R&D project manager to make suboptimal decisions.

Budgeting for all types of projects, including R&D projects, must begin with a solid foundation based upon effective planning. Enterprise project management templates are mandatory for this to be done effectively on a repetitive basis. Only when there are adequate project plans with well-defined objectives can R&D project managers improve their chances of success. These projects, once approved, generate the need for an effective method of budgeting, feedback, and control to verify that the work is progressing according to the strategic direction. Unlike other projects, every R&D project can be viewed as part of the executive's strategic plan. However, the control function is critical. Executives must be willing to pull the plug on R&D projects that do not appear to be going well and redirect the funds to projects where the chances of success are greater or projects that can deliver greater business value. This mandates that executives and stakeholders work much closer with the project managers throughout the life of the project. This is an important characteristic of PM 2.0 as mentioned in Chapter 1.

The R&D project management function, regardless of its position within the organizational structure of the firm, can be viewed as a strategic planning system with an input of money and an output of products. Furthermore, as with most systems, a feedback mechanism must exist such that the flow of money can be diverted from less promising projects to those with greater potential. This implies that the R&D systems must be continuously monitored, perhaps more so than the other systems. It also implies that both the stakeholders and the executives monitoring the system have a reasonable understanding of project management and hopefully R&D. Once again, this justifies the need for PMO involvement to ensure that the monitoring process goes smoothly and creates minimum disturbance to the R&D projects.

9.2 PRODUCT PORTFOLIO ANALYSIS

When a corporation develops its strategic plan for R&D, the plan must be aligned with the corporation's strategic business unit and portfolio of products or services. R&D projects are selected to enhance the portfolio of products or services by either adding new products or continuous improvements to existing products. A corporation should have a balanced mix of products, and the projects selected are designed to support each type of product or product line.

The same holds true when performing R&D for external clients. In Chapters 1 and 2, we discussed that project managers working in a PM 2.0 environment must fully understand the client's business model, and this may require an understanding of the client's business strategy. Simply stated, the project must be aligned to the client's business strategy, and therefore knowledge about the client's business is essential.

There are several ways by which a portfolio of products can be represented. The three most common methods are the Boston Consulting Group (BCG) model, the General Electric (GE) model, and the product life-cycle model.[1] Each model brings with it advantages and disadvantages, but all three models can be used to assist in the selection of R&D projects.

The first model is the BCG model and is shown in Figure 9-2.

In the BCG model, the product mix consists of stars, cash cows, dogs, and question marks, as shown in Figure 9-2. For simplicity sake, we can classify each product as follows:

Stars: These products represent the company's best opportunity for long-term growth and profitability. These stars are supported by R&D projects to enhance quality, find other potential uses for the products, spin-offs, and continuous improvement efforts. The focus of the R&D projects includes increasing market share and finding new uses/customers for the products. Because of the heavy investment in R&D projects, stars are often cash users rather than cash generators. However, most stars are generally low-risk investments in R&D and most of the R&D investment funds eventually get recovered.

---

[1] Most textbooks on strategic planning and strategic management cover these models in detail. The intent here is simply to show some of the models that are used to help select R&D projects.

STRATEGIC SELECTION OF PROJECTS

FIGURE 9-2 **BCG model**

Question Marks: These products, often called problem children, become either stars or dogs. To convert a question mark into a star requires a very heavy investment in R&D with the focus on creating product features that will enhance new market penetration and increase market share. R&D investment in question mark products is risky because, if the product becomes a dog rather than a star, the funds invested in R&D will never be recovered.

Cash Cows: These products have a relatively high share of the market but the growth opportunities are limited or declining. The focus of R&D is getting the cash cow to live as long as possible for milking purposes. Typical R&D projects would include quality improvements, upgrades, and enhancements.

Dogs: These products are either liquidated or divested because they consume a great deal of effort with very limited return. For a dog to become a question mark or star, a significant technical breakthrough would be required. This could necessitate a heavy investment in basic or applied research along with relatively unknown time frames. Significant luck may be required to recover R&D costs related to dogs.

Some products generate cash, while others use cash. Obviously, in order to manage the company in the short term, cash cows are required. On the other hand, in both the short and long term, a company needs stars coming into the marketplace and in development for future growth. It is doubtful that a company would want to invest very heavily in R&D projects for the development of new products in a declining mature market. In this example, a company may choose to emphasize R&D cost reduction projects associated with existing products.

It is important for the project manager to understand in which quadrant the project resides. Making either project or business decisions must be based upon the quadrant. The type of R&D and the accompanying business value expected are also based upon the quadrant. The need for PM 2.0 project managers to have more of a business background or business knowledge should now be apparent.

Similar to the BCG model is the GE model shown in Figure 9-3. The GE model has a grid of nine boxes whereas the BCG model had only four boxes.

The selection of R&D projects for each box in the grid is based upon the desired competitive position and market attractiveness. Items that are looked at in these areas are shown in Figures 9-4 and 9-5.

A company must develop product line strategies that are linked to its R&D project selection strategies. A balanced product portfolio means new products must be under development or in their growth phase to replace those in maturity or already in decline. A continuous product improvement program is necessary to remain ahead of the competition. This is illustrated in Figure 9-6, which shows the life-cycle model. In Figure 9-6, the circles represent the size of the industry and the pie-shaped wedge illustrates the product's market share. Using Figure 9-6, we can select R&D projects related to a product's life-cycle phase.

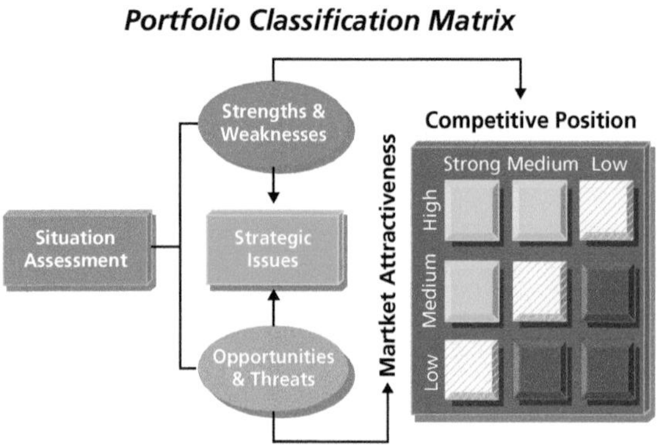

Portfolio Classification Matrix

FIGURE 9-3 GE model

FIGURE 9-4 Competitive position

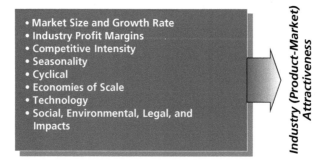

FIGURE 9-5 Market attractiveness

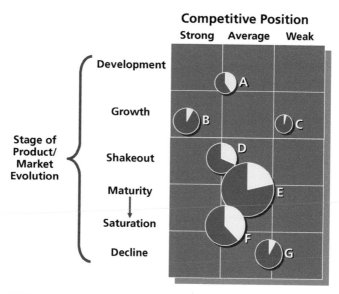

FIGURE 9-6   Life-cycle model

9.3 MARKETING INVOLVEMENT WITH R&D PROJECT MANAGERS

From the previous section, it should be obvious that marketing often provides the greatest input in identifying the critical variables for R&D project selection. Marketing involvement in R&D project management using PM 2.0 characteristics includes identification of:

- Market attractiveness for each product/product line
- Relative market share (and trend)
- Sales growth rate (and trend)
- Current business strength
- Forecasted market share trend
- Recommended investment strategy

For the selection of an R&D project designed to create new features or a new product, additional factors include marketing's policy toward first to market, follow the leader, applications engineering, and "me too." It should be evident that R&D project managers must have a good knowledge about the business in general, especially a window into the future.

First to Market

This risky but potential rewarding strategy has a number of important ramifications throughout the business:

- A research-intensive effort, supported by major development resources
- Close downstream coupling in product planning and moderately close coupling thereafter
- High proximity to the state of the art
- High R&D investment ratio
- A high risk of failure for individual products

Taken together, they outline a clear philosophy of business. The company must recruit and retain outstanding technical personnel who can win leadership positions in the industry. The company must see that these technical R&D project managers are in close and useful communication with marketing planners to identify potentially profitable markets. The company must often risk large investments of time and money in technical and market development without any immediate return. It must be able to absorb mistakes, withdraw, and recoup without losing its position in other product lines. As the nature of the market clarifies, initial plans must quickly be modified and approximation refined into precision.

Perhaps most important, top management must be able to make critical judgments of timing, balancing the improved product development stemming from a delayed introduction against the risk of being second into the market. Such a company must have more than its share of long-range thinkers who can confidently assess market and competitive trends in their earliest stages and plan with both confidence and flexibility. However, project management involvement is still needed to assess whether the desires by management can actually be accomplished and the associated delivery risks.

Follow the Leader

This marketing strategy implies:

- Development-intensive technical effort
- Moderate competence across the spectrum of relevant technologies
- Exceptionally rapid response time in product development and marketing on the basis of finished research
- High downstream coupling of R&D with marketing and manufacturing
- Superior competitive intelligence
- R&D personnel that are knowledgeable about the competitors' products

The company that follows this strategy is—or should be—an organization that gets things done. R&D project management must have a reasonably good success record. This will be essential in a PM 2.0 environment. It uses many interfunctional techniques, responds rapidly to change, and often seems to be in a perpetual fire drill. It has few scientists on its payroll but some of the best development engineers available. Its senior executives are constantly concerned with maintaining the right balance of strengths among the technical, marketing, and manufacturing functions so that the company can respond effectively to the leader's moves in any of these three areas.

Application Engineering

This strategy requires:

- Substantial product design and engineering resources but no research and little real development
- Ready access to product users within customer companies
- Technically perceptive salesmen and sales engineers who work closely with product designers
- Good product line control to prevent costly proliferation
- Considerable cost consciousness in deciding what applications to develop
- An efficiency-oriented manufacturing organization
- A flair for minimizing development and manufacturing cost by using the same parts or elements in many different applications.

The applications engineering strategy tends to avoid innovative efforts in the interest of the economy. Planning is precise, assignments are clear, and new technology is introduced cautiously, well behind the economic state of the art. Return-on-investment and cash flow calculations are standard practice, and the entire management is profit oriented.

"Me Too"

This strategy, which has flourished in the past decade as never before, is distinguished by:

- No research or development
- Strong manufacturing function, dominating product design
- Strong price and delivery performance
- Ability to copy new designs quickly, modifying them only to reduce production costs

Competing on price, taking a low margin, but avoiding all development expense, a company that has adopted this strategy can wreak havoc with competitors following the first-to-market or follow-the-leader strategies. This is because the me-too strategy effectively pursued shortens the profitable period after introduction when the leaders' margins are most substantial. The me-too strategy requires a low-overhead approach to manufacturing and administration and a direct hard sell on price and delivery to the customer. It does not require any technical enthusiasm, nor does it aim to generate any.

Business environments are highly dynamic. As such, the company may find it necessary to change strategies in midstream. This can play havoc with R&D project management. This is just another reason why R&D project managers in a PM 2.0 environment must have business knowledge. Factors that can lead to rapid changes in R&D projects are changes in technology, new product introduction by a competitor, new competitors entering the marketplace, changes in cash flow resulting in a cutback in R&D funding, and even the whims of management.

9.4 PRODUCT LIFE CYCLES

The length of the product life cycles has a bearing on how much risk the company should take, and this risk is then transferred to the R&D project managers. Long product life cycles may generate sufficient cash flow and lead time for new product introduction. R&D project managers may be under very little pressure to develop new products in a short time frame.

Unfortunately, many companies put the emphasis in the wrong place. As the product life cycle grows, emphasis shifts from a competitive position, based on product performance, to product variations and ultimately lower costs due to learning curve effects. Instead of having the R&D project managers focus on the development of new products, emphasis is on the continuation of existing products perhaps through enhancements.

Many companies prefer taking the low risk (costwise) of developing "line extension" or "flanker" products, which are simply the same product in a different form. Needless to say, bringing out something new is a risky business, and some companies simply prefer to develop cheap imitations. Many small companies thrive on short product life cycles, even in fragmented markets. As an example, consider that if the top executive of a small engineering company is also the founding genius and top scientist, new products are developed quickly and move rapidly from R&D to the marketplace. This company has learned to cope well with short product life cycles.

9.5 R&D PROJECT PLANNING ACCORDING TO MARKET SHARE

Whenever market share analysis is combined with market growth analysis, the executive is provided with an excellent tool to determine whether there exists an investment opportunity, a source of cash, or an item that should be removed from service, such as was shown previously with the BCG model for product portfolio analysis. The general terminology for these elements of the market share/market growth matrix or BCG model is cash cows, stars, dogs, and either question marks or problem children. In portfolio planning, each of these, as discussed previously, has a bearing upon the direction of the R&D thrust, which is to maintain market share, build market share, harvest, or simply withdraw.

Maintain Market Share: This strategy represents a stable market and is ideal for stars or cash cows. In this case, the accompanying R&D strategy should stress defensive R&D and applications engineering rather than diversification.

Build Market Share: This strategy is ideal for selected stars and problem children. The R&D strategy to support this would include methods for lowering production costs, improving quality, and applications engineering. This build-market-share strategy can also be used for dogs, provided that the company sees a technological breakthrough that will drastically increase market share and perhaps some degree of patent protection that will guarantee a profitable life cycle.

Harvest Market Share; This strategy is used for a cash cow, where the funds are needed for other activities as R&D. A reasonable R&D strategy here would be to improve the quality or lower the cost.

Withdraw Market Share: This strategy is used with dogs, where the troublesome product has a very low market share and is marginally profitable or operating at a loss. The R&D strategy, if employed at all, should be to look for spin-offs, specialized applications if profitable, or minimal defensive R&D to support future activities that may need this technology as a base.

9.6 CLASSIFICATION OF R&D PROJECTS

From the previous sections, it should be obvious that there are numerous ways for R&D projects to be classified based upon the relationship between R&D and marketing. Figure 9-7 shows a simple way to classify R&D projects for PMO involvement.

The amount of involvement varies based upon the risks, strategic importance, organizational maturity in project management, and experience level of the project managers.

Another simple approach is to classify R&D projects into seven major categories:

1. **Grass Roots Projects:** This type of project can be simply an idea with as few as one or two good data points. Grass roots projects are funded with "seed money," which is a small sum of money usually under the control of the R&D project manager. The purpose of the seed money is to see if the grass roots project is feasible enough to be developed into a full-blown, well-funded R&D project to be further incorporated into the strategic plan.

2. **Bootlegged Projects:** This type of project is one in which funding does not exist either because the selection team did not consider this project worthy of funding or because funding had been terminated (or ran out) and funding renewal was

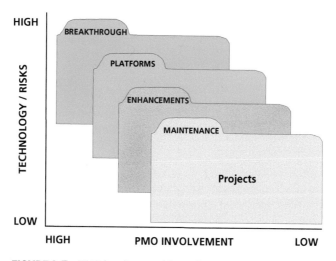

FIGURE 9-7 PMO involvement in projects

not considered appropriate. In either event, a bootlegged project is done on the sly, using another project's budgeted charge numbers. Bootlegged projects run the complete spectrum for conceptual ideas to terminated, well-defined activities.

3. **Basic Research Projects:** This category may include the grass roots and bootlegged projects. Basic research activities are designed to expand knowledge in a specific scientific area or to improve the state of the art. These types of projects do not generally result in products that can be directly sold by marketing and, as a result, require special handling. Instead, they create intellectual property to be used on future projects.

4. **Applied Research Projects:** The applied research project is an extension or follow-on to the basic research project and explores direct application of a given body of knowledge. These types of projects hopefully result in marketable products, product improvements, or new applications for existing products.

5. **Advanced Development Projects:** These types of activities follow the applied research or exploratory development projects with the intent of producing full-scale prototypes supported by experimental testing.

6. **Full-Scale Development Projects:** This activity includes, complete, working, drawing design of the product together with a detailed bill of materials, exact vendor quotes, and specification development. This type of R&D activity requires strong manufacturing involvement.

7. **Production Support Projects:** The seventh and final category is the production support R&D project and can include either applications engineering to find better uses of this product for a customer or internal operations support to investigate limitations and feasibility of a given system with hopes of modification or redesign. Projects designed to find ways of lowering production costs or improving product quality are examples of internal production support projects.

9.7 RESEARCH VERSUS DEVELOPMENT

Although most people consider R&D project management as a total entity, there are critical differences between research and development. Some project managers are excellent in managing research projects while others excel in development projects. Basic differences include:

Specifications: Researchers generally function with weak specifications because of the freedom to invent, whereas development personnel are paid not to create new alternatives but to reduce available alternatives to one hopefully simple solution available for implementation.

Resources: Generally, more resources are needed for development work rather pure research. This generates a greater need for structured supervision, whereas research is often conducted in a campuslike work environment.

Scheduling: Researchers prefer very loose schedules with the freedom to go off on tangents, whereas developmental schedules are more rigid. Research schedules identify parallel activities whereas in development scheduled activities are sequential.

Engineering Changes: In the research stages of a project, engineering changes, specification changes, and engineering redirection (even if simply caused by the personal whims of management) may have a minor cost impact compared to the same changes occurring in the development stage.

For simplicity's sake, however, we will assume that R&D is a single entity.

9.8 R&D RATIO

A company must determine the balance or ratio of basic research, applied research, and product development with project managers specially trained in each area. Basic research may or may not respond directly to a specific problem or need, but it is selected and directed into those fields where advances will have a major impact on the company's future core businesses.

Applied research is the next step in using a technology to accomplish business objectives that may include processes, cost reduction, and so on. Product development uses all technologies available to it to develop a product that is consistent with the direction of the company.

An industrial company needs to determine the ratio of the above-mentioned areas when determining its strategy for short- and long-term decisions. Basic research is generally a long-term commitment that must be made and driven by top management. Marketing, sales, and manufacturing do not have the incentive to sponsor applied research and more product development research for short-term programs. Therefore, it is the responsibility of top management to provide direction to the R&D effort within the corporation.

Functional groups must work closely with the PMO in the strategies selection of projects. Each group brings to the table different ideas that usually favor their own functional area.

Manufacturing and Sales

These two groups must be included in the portfolio selection of projects for R&D. This is particularly true for the development of new products and technologies. It is essential to know if manufacturing has the capability to make the product using existing manufacturing facilities and equipment. Will the existing manufacturing plants have sufficient capacity to meet demand? Will they be able to manufacture the product and be cost effective? If new equipment and manufacturing plants are needed, this information needs to be factored into the overall plan so they will be ready when the new product is ready to be launched. Just as important is the sales force. Is the present sales force adequate? Adequacy must be evaluated in terms of numbers, training, location, and so on. Will the new product require different selling skills than the company's present product line? Another factor that must be evaluated is the possible reduction in sales force due to a new product. What adverse effect might that have on the morale of the sales group? What are the behavioral ramifications of such a move?

Human Behavior

One of the key factors in project management is the ability to communicate effectively with a great deal of emphasis placed on teamwork, interaction between groups, and knowing your customer. Obviously, the more top management and project management understand human behavior, the more they can control productivity and the management of limited human resources.

One of the many problems associated with the R&D project management is ownership. Top management may feel that they have the need and the right to constantly control a project. With PM 2.0, management must relinquish control in order to allow R&D project managers to inject the degree of creativity needed to make the project successful. On the other hand, the individual or team within R&D who is working on the project also needs to relinquish control once the project is ready to be released to manufacturing and marketing. Unlike other forms of project management, R&D project managers may not be involved throughout the complete life cycle of the project unless downstream changes are needed.

9.9 OFFENSIVE-VERSUS-DEFENSIVE R&D

Should a company direct its resources toward offensive or defensive R&D? Offensive R&D is product R&D, whereas defensive R&D is process R&D. In offensive R&D, the intent is to penetrate a new market as quickly as possible, replace an existing product, or simply satisfy a particular customer's need. Offensive R&D stresses a first-to-market approach. The ultimate decision can be related to some of the models discussed previously, such as the GE nine-square grid shown in Figure 9-8.

Defensive R&D, on the other hand, is used to either lengthen the product life cycle or protect existing product lines from serious competitive pressures. Defensive R&D is

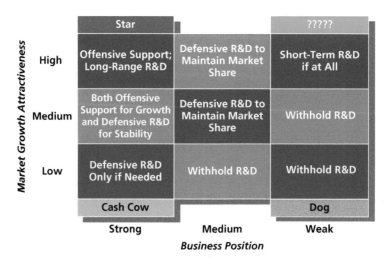

FIGURE 9-8   Positioning the R&D strategy

also employed in situations where the company has a successful product line and fears that the introduction of a new technology at this time may jeopardize existing profits. Defensive R&D concentrates on minor improvements rather than major discoveries and, as a result, requires less funding. Today, with the high cost of money, companies are concentrating on minor product improvements, such as style, and introduce the product into the marketplace as a new, improved version when, in fact, it is simply the original product slightly changed. This approach has been used successfully by the Japanese in copying someone else's successful product, improving the quality, changing the style, and introducing it into the marketplace. A big advantage to the Japanese approach is that the product can be sold at a lower cost, because the selling price does not have to include recovery of expensive R&D costs.

Defensive R&D is a necessity for those organizations that must support existing products and hopefully extend the product's life span. A firm's strategic posture in the marketplace is, therefore, not restricted solely to new-product introduction. Companies must find the proper technological balance between offensive and defensive R&D.

9.10 MODELING R&D PLANNING FUNCTION

Schematic modeling of the R&D project management requires an understanding of how R&D fits into the total strategic plan and the R&D functional strategy. Figure 9-9 illustrates the integration of R&D into the total strategic planning function. Once the business is defined, together with an environmental analysis of strengths, weaknesses, opportunities, and threats, the corporate goals and objectives are defined. Unfortunately, the definition of the strategic goals and objectives is usually made in financial

FIGURE 9-9   **R&D strategic planning process**

terms or through the product/market element. This type of definition implies a critical assumption: R&D can and will develop the new products or product improvements within the required specifications in order to meet target goals and objectives. Unfortunately, many companies have not realized the importance of soliciting R&D input into the objective-setting stage and, therefore, treat R&D simply as a service organization. Once the objectives are set, marketing will identify the products and approach (tactics) to achieve the strategies. Here, again, R&D may be treated as a service organization.

The PMO, along with the appropriate senior managers, will manage the portfolio selection process for the R&D projects. It is not uncommon for either R&D selection process to be controlled by marketing or for the entire R&D budget to be part of the marketing budget. The reason for this is because marketing wants to be sure that it can sell successfully what R&D produces.

In mature organizations, however, R&D personnel are allowed to express their concerns over the feasibility of the goals and objectives and of the probability of successfully achieving the R&D objectives. In such a case, there exists a feedback loop from project selection to objective setting, as shown in Figure 9-9.

The box in Figure 9-9 entitled "Support for New Products" requires that the R&D project selection process account for the management of innovation and entrepreneurship and can be modeled as shown in Figure 9-10. Not all companies have entrepreneurship strategies because of a slow and tedious permeation process into the corporate culture. The successful companies consider entrepreneurship as "business" in itself and marry it to the mainstream of the company. In this case, R&D project managers are expected to possess some degree of entrepreneurship skills.

Figure 9-9 also shows that successful integration of R&D into the strategic plan requires that R&D understand the firm's production process, distribution process,

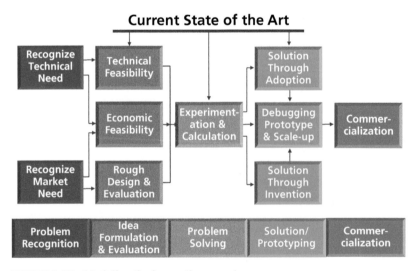

FIGURE 9-10   **Modeling the innovation process**

market research, and market distribution channels. This requires that R&D understand marketing's decision to introduce a new product by being first to market, second to market, application engineering, or me too (i.e., copycat).

When a company has perceived a strategic need to enter a new market, increase growth, or improve an existing product, the company is faced with the problem of how to acquire the technical skills necessary for integration in the strategic plan. The alternatives are:

- R&D with existing resources that have the technical capability
- R&D with existing resources through internal technical training
- R&D through newly hired employees
- R&D through consultants
- Acquisition of a company with the required technology
- Joint ventures
- Buying technology through licensing

As described previously, functional strategies may be performed independently up to the point of integration into the total plan. The functional R&D strategy was shown in Figure 9-9. The environmental analysis box can also be accomplished entirely by R&D project managers to stay abreast of the state of the art and what the competition is doing. It is also important to note that the termination of each project should result in an updating of the strategic plan for R&D.

There are basic differences in strategic planning for R&D as a result of short and long product life cycles. For short product life cycles, R&D project managers must be willing to respond rapidly, especially if the environment is ever changing. Adaptability to short product life cycles is characteristic of flat organizational structures with a wide span of control. Because decision making must be quick in short product life cycles, organizational coupling must be high between marketing, R&D, and manufacturing. Weak coupling can result in the late introduction of new products into the marketplace.

The shorter the product life cycle, the greater will be the involvement of senior management. Strategic planning at the strategic business unit (SBU) level may be cumbersome with short product life cycles. The shorter the product life cycle, the greater number of new products needed to sustain a reasonable growth. As a result, shorter product life cycles have a greater need for superior R&D talent.

As a final note, the strategic marketing approach to the product life cycle can vary based upon the size of the company. Small companies that compete in short product life-cycle markets must be first to market to reap profits. In short product life cycles, large companies can commit vast resource to take advantage of experience curves, thus creating a barrier to entry for smaller companies that try to employ a follow-the-leader approach.

Templates can be developed for various components of the modeling function. As an example, Figure 9-11 shows a way of establishing risks for various types of R&D projects.

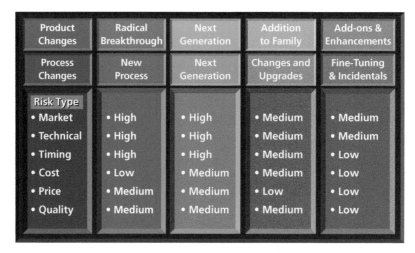

Product Changes	Radical Breakthrough	Next Generation	Addition to Family	Add-ons & Enhancements
Process Changes	New Process	Next Generation	Changes and Upgrades	Fine-Tuning & Incidentals

Risk Type				
• Market	• High	• High	• Medium	• Medium
• Technical	• High	• High	• Medium	• Medium
• Timing	• High	• High	• Medium	• Low
• Cost	• Low	• Medium	• Medium	• Low
• Price	• Medium	• Medium	• Low	• Low
• Quality	• Medium	• Medium	• Medium	• Low

FIGURE 9-11   Risk intensity

9.11 PRIORITY SETTING

Priorities create colossal management headaches for the R&D project manager because R&D projects are usually prioritized on a different list from all other projects. Functional managers must now supply resources according to two priority lists. Unfortunately, the R&D priority list is usually not given proper attention.

As an example of this, the director of R&D of a Fortune 25 corporation made the following remarks:

> Each of our operating divisions has their own R&D projects and priorities. Last year, corporate R&D had a very high R&D project geared toward cost improvement in the manufacturing areas. Our priorities were based upon the short-run requirements. Unfortunately, the operating division that had to supply resources to our project felt that the benefits would not be received until the long run and, therefore, placed support for our project low on their priority list.

Communication of priorities is often a problem in the R&D area. Setting of priorities on the divisional level may not be passed down to the departmental level, and vice versa. We must have early feedback of priorities so that functional managers can make their own plans.

Working with Marketing

In most organizations, either R&D drives marketing or marketing drives R&D. The latter is more common. Well-managed organizations maintain a proper balance between marketing and R&D. Marketing-driven organizations can create havoc, especially if marketing continuously requests information faster than R&D can deliver it and if bootleg R&D is eliminated. In this case, all R&D funding comes out of the marketing budget.

In order to stimulate creativity, R&D should have control over at least a portion of its own budget. This is a necessity, because not all R&D activities are designed to benefit marketing. Some activities are simply to improve technology or create a new way of doing business.

Marketing support, if needed, should be available to all R&D projects regardless of whether they originate in marketing or R&D. An R&D project manager at a major food manufacturer made the following remarks:

> A few years ago, one of our R&D people came up with an idea and I was assigned as the project manager. When the project was completed, we had developed a new product, ready for market introduction and testing. Unfortunately, R&D does not maintain funds for the market testing of a new product. The funds come out of marketing. Our marketing people did not understand the product and placed it low on their priority list. We in R&D tried to talk to them. They were reluctant to test the new product because the project was our idea. Marketing lives in its own little world. To make a long story short, last year one of our competitors introduced the same product into the marketplace. Now, instead of being the leader, we are playing catch-up. I know R&D project managers are not trained in market testing, but what if marketing refuses to support R&D-conceived projects? What can we do?

Several organizations today have R&D project managers reporting directly to a new business group, business development group, or marketing. Engineering-oriented R&D project managers continuously voice displeasure at being evaluated for promotion by someone in marketing that really may not understand the technical difficulties in managing an R&D project. Yet, executives have valid arguments for this arrangement, asserting that these high-technology R&D project managers are so in love with their projects that they do not know how and when to cancel a project. Marketing executives contend that projects should be canceled when:

- Costs become excessive, causing product cost to be noncompetitive
- Return on investment will occur too late
- Competition is too stiff and not worth the risk

Of course, the question arises, "Should marketing have a vote in the cancellation of each R&D project or only those that are marketing driven?" Some organizations cancel projects upon consensus of the project team.

Companies turn out to be R&D dominated, marketing dominated, or balanced. In R&D-dominated companies, the R&D staff researches fundamental problems, looks for major breakthroughs, and strives for technical perfection in product development. R&D expenditures are high, and the new-product success rate tends to be low, although R&D occasionally comes up with major new products.

In marketing-dominated companies, the R&D staff designs products for specific market needs, much of it involving product modification and the application of existing technologies. A high ratio of new products eventually succeed, but they represent mainly product modifications with relatively short product lives.

A balanced R&D/marketing company is one in which effective organizational relationships have been worked out between R&D and marketing to share responsibility for successful market orientation innovations. The R&D staff takes responsibility not for invention alone but for successful innovation. The marketing staff takes responsibility not for new sales features alone but also for helping identify new ways to satisfy needs. R&D/marketing cooperation is facilitated in several ways:

- Joint seminars are sponsored to build understanding and respect for each other's goals, working styles, and problems.
- Each new project is assigned to an R&D person and a marketing person who work together through the life of the project.
- R&D and marketing personnel are interchanged so that they have a chance to experience each other's work situations (some R&D people may travel with the sales force, while some marketing people might hang around the lab for a short time).
- Higher management, following a clear procedure, works out conflicts.

9.12 CONTRACT R&D

Contract R&D is another form of strategic planning for R&D project management that can be used with any of the seven classifications for projects. There are different reasons for conducting contract R&D, depending on whether you are the customer or the contractor. Customers subcontract out R&D because they may not have the necessary in-house technical skills; have in-house skills, but the resources are committed to higher priority activities; and/or may have the available talent, but external sources have superior talent and may be able to produce the desired results in less time and for less money.

From a subcontractor's point of view, contract R&D project management is a way to develop new technologies at someone else's expense. Subcontractors view contract R&D as a way to:

- Minimize the internal cost of supporting R&D personnel
- Develop new technologies to penetrate new markets/products
- Develop new technologies to support existing market/products
- Maintain technical leadership
- Improve resource utilization by balancing workloads
- Maintain customer goodwill
- Look for spin-offs on existing products

There also exist disadvantages to contract R&D from the customer's point of view:

- How dependent should I become on a subcontractor to produce the desired results within time and cost?
- What criteria are used to evaluate subcontractors?
- What type of communication network should be established?
- How do I know if the subcontractor is being honest with me?
- If trade-offs are needed, how will decisions be made?
- Who controls patent rights resulting from the research under contract?
- Will project failures impact the strategic planning process?

From the subcontractor's point of view:

- What influence will the customer try to exercise over my personnel?
- Will project success generate follow-on work? Will project success enhance goodwill image? Will project failure result in a loss of future revenue?

9.13 NONDISCLOSURE AGREEMENTS, SECRECY AGREEMENTS, AND CONFIDENTIALITY AGREEMENTS

In the course of doing business, few companies can develop and market a new product without some help from outside their company. R&D project managers, perhaps more so than any other type of project manager, must have some knowledge about nondisclosure agreements, secrecy agreements, and confidentiality agreements. When it is necessary to secure outside help, it is essential to protect the proprietary nature of the information being transmitted to the outside party. In order to do this, an agreement is drawn up between the two parties and signed by a corporate office from each firm.

Top management must develop a policy on how to handle the transfer of confidential information regarding technological developments to outside sources involved in the project. Generally speaking, there are two types of agreements: one way and mutual. The one-way agreement is exactly what the name implies in that the company with the confidential information is transferring that information to the second party with nothing coming back. The mutual agreement calls for the transfer of confidential information between both parties.

9.14 GOVERNMENT INFLUENCE

Foreign and domestic governments play a significant role in the strategic-planning process for R&D and can therefore have significant influence over the way that R&D projects are managed. The laws and policies set forth by government can encourage or discourage R&D. This effect can be direct or indirect. The government may have tax incentives that will foster a climate for R&D to flourish. Government can also impose regulations or standards that will encourage the development of new products to meet those standards. The behavior and posture of foreign governments can influence licensing agreements, the competitive edge on new products, the ability to market new products, and so on.

The following list is but a sampling of how government can control and influence industrial R&D.

- Fiscal and monetary policies
- International operations and control
- Technology transfer restrictions
- Patents
- Policy impact on technological corporations
- Taxes; monetary flow restrictions
- Labor/management relations
- Risk
- Regulation
- Sponsor of technological advance, with corporate involvement

Joseph P. Martino identifies additional concerns that may be more of a political nature[2]:

- Regulations regarding product safety to users
- Regulation of effectiveness

---

[2] J. P. Martino, *R&D Project Selection*, Wiley, New York, 1995, p. 107.

- Safety regulation of the industry using the product or process
- Economic regulation of the industry using the product or process
- Regulations regarding workplace safety during manufacture
- Regulations regarding environmental hazards
- Regulations regarding disposability or recyclability

Contract research, licensing, joint ventures, acquisitions, and the luxury of hiring additional personnel are taken for granted in the United States. Foreign countries may not have these luxuries, and additional classification may be needed, usually by the level of technology. According to one foreign country, the following levels are used:

- **Level I:** Technology exists within the company.
- **Level II:** Technology can be purchased from companies within the country.
- **Level III:** Technology can be purchased from outside the country.
- **Level IV:** Technology must be researched in other countries and brought back into the parent country.

Because a great many foreign countries fall into levels III and IV, several foreign corporations have established employee sabbatical funds. Each month the company withholds 3% of the employee's salary and matches this with 7% of company funds. Every five or six years, each participating employee is allowed to study abroad to bring technical expertise back into the country. The employee draws his or her full salary while on sabbatical in addition to the sabbatical fund.

For strategic R&D planning, this type of sabbatical leave creates a gap in the organization. Management can delay a sabbatical leave for an employee for one year only. What happens if the employee is in a strategic position? What if the employee is working on a critical project? What if the employee is the only person with the needed skills in a specific discipline? Who replaces the employee? Where do we put the employee upon his or her return to the organization? What happens if the employee's previous management slot is no longer vacant? Obviously, these questions have serious impact on the strategic-planning process.

9.15 SOURCES OF IDEAS

What are the sources for ideas?

Unlike other types of planning, strategic R&D project planning must be willing to solicit ideas from the depths of the organization. Successful companies with a reputation for continuous new-product introduction have new-product development teams that operate in a relatively unstructured environment to obtain the best possible ideas. Some companies go so far as to develop idea inventories, idea banks, and idea clearinghouses.

These idea sessions are brainstorming sessions and not intended for problem solving. If properly structured, the meeting will have an atmosphere of free expression and creative thinking, an ideal technique for stimulating ideas. Arguments against brainstorming sessions include no rewards for creators, attack of only superficial problems, possibility of potentially good ideas coming out prematurely and being disregarded, and lack of consideration for those individuals who are more creative by themselves.

Principles that can be used in brainstorming sessions include:

- Select personnel from a variety of levels; avoid those responsible for implementation.
- Allow people to decline assignments.
- Avoid evaluation and criticism of ideas.
- Provide credit recognition and/or rewards for contributors.
- Limit session to 60 min.

Ideas are not merely limited to internal sources. There are several external sources of new-product ideas, such as:

- Customers
- Competitors
- Suppliers
- Purchase of technologies
- Licensing of technologies
- Unsolicited ideas from customers or others
- Private inventors
- Acquisitions
- Trade fairs
- Technology fairs
- Private data banks
- Technical journals
- Trade journals
- Government-funded research programs
- Government innovation/technology transfer programs
- Government agencies

Innovation is expensive and the more ideas you have, the greater the chance for a commercially successful product to emerge. In a study of new-product activities of several hundred companies in all industries, Booz, Allen & Hamilton defined the new-product evolution process as the time it takes to bring a product to commercial existence.[3] This process began with company objectives, which included fields of product interest, goals, and growth plans, and ended with, hopefully, a successful product. The more specific these objectives were defined, the greater the guidance given to the new-product program. This process was broken down into six manageable, fairly clear sequential stages:

Exploration: The search for product ideas to meet company objectives.

Screening: A quick analysis to determine which ideas were pertinent and merit more detailed study.

Business Analysis: The expansion of the idea, through creative analysis, into a concrete business recommendation, including product features, financial analysis, risk analysis, market assessment, and a program for the product.

---

[3] For a more detailed analysis of these six categories leading to commercialization, see *Management of New Products,* Booz, Allen & Hamilton, 1984. This report does an excellent job describing life-cycle phases to get to commercialization and the complexities of each phase.

Development: Turning the idea on paper into a product in hand, demonstrable and producible. This stage focuses on R&D and the inventive capability of the firm. When unanticipated problems arise, new solutions and trade-offs are sought. In many instances, the obstacles are so great that a solution cannot be found, and work is terminated or deferred.

Testing: The technical and commercial experiments necessary to verify earlier technical and business judgments.

Commercialization: Launching the product in full-scale production and sale; committing the company's reputation and resources.

In the Booz, Allen & Hamilton study, the new-product process was characterized by a decay curve for ideas, as shown in Figure 9-12. This showed a progressive rejection of ideas or projects by stages in the product evolution process. Although the rate of rejection varied between industries and companies, the general shape of the decay curve is typical. It generally takes close to 60 ideas to yield one successful new product.

The process of new-product evolution involves a series of management decisions. Each stage is progressively more expensive, as measured in expenditures of both time and money. Figure 9-13 shows the rate at which expense dollars are spent as time accumulates for the average project within a sample of leading companies. This information was based on an all-industry average and is therefore useful in understanding the typical industrial new-product process. It is significant to note that the majority of capital expenditures are concentrated in the last three stages of evolution. It is therefore very important to do a better job of screening for business and financial analysis. This will help eliminate ideas of limited potential before they reach the more expensive stages of evolution.

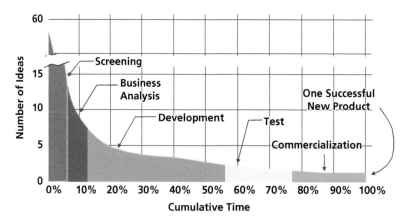

FIGURE 9-12   Mortality of new ideas

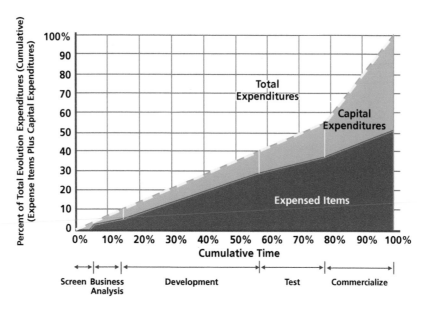

FIGURE 9-13   Cumulative expenditures and time

9.16 ECONOMIC EVALUATION OF PROJECTS

There are several methods available for the economic evaluation of a single R&D project. According to Martino, these methods include[4]:

- Ranking methods
  - Pairwise comparison
  - Scoring models
  - Analytic hierarchy procedure
- Economic methods
  - Net present value
  - Internal rate of return
  - Cash flow payback
  - Expected value
- Portfolio optimization methods
  - Mathematical programming
  - Cluster analysis
  - Simulation
  - Sensitivity analysis

---

[4] J. P. Martino, p. 192.

- Ad hoc methods
  - Profiles
  - Interactive methods
  - Cognitive modeling
- Multistage decisions
  - Decision theory

Martino also identifies factors that can be used for evaluating R&D projects one against another[5]:

- Factors that can be included
  - Cost
  - Payoff
  - Probability of technical success
  - Probability of market success
  - Market size
  - Market share
  - Availability of required staff
  - Degree of organizational commitment
  - Strategic positioning for project
  - Degree of competition
  - Favorability of regulatory environment
- Special input requirements
  - Requires precise cash flow information
  - Requires precise life-cycle information
  - Requires probability of technical success
  - Requires probability of market success
- Special features
  - Considers resource dependencies
  - Considers budget constraints
  - Considers technical interactions
  - Considers market interactions
  - Incorporates program considerations
  - Can be used for large numbers of projects
  - Allows comparison with other investments
  - Suited for research stage
  - Suited for development stage

Typical rating models are shown in Figure 9-14, Table 9-1, and Figure 9-15.[6] These models can be used for both strategic selection and prioritization.

---

[5] J. P. Martino, p. 193.
[6] W. Souder, *Project Selection and Economic Appraisal*, Van Nostrand Reinhold, New York, 1984, pp. 66–69.

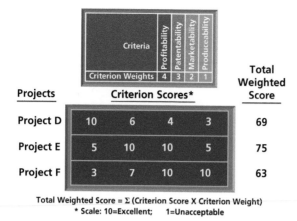

Projects	Criterion Scores*				Total Weighted Score
Project D	10	6	4	3	69
Project E	5	10	10	5	75
Project F	3	7	10	10	63

Total Weighted Score = Σ (Criterion Score X Criterion Weight)
* Scale: 10=Excellent; 1=Unacceptable

FIGURE 9-14   Illustration of a scoring model

TABLE 9-1 Illustration of a Checklist for Three Projects

	Criteria									
	Profitability			Marketability			Success Likelihood			
Projects	3	2	1	3	2	1	3	2	1	Total Score
Project A	√				√			√		7
Project B		√		√					√	6
Project C			√		√				√	3

9.17 R&D PROJECT READJUSTMENTS

Many R&D projects are managed by overly optimistic prima donnas who truly believe that they can develop any type of product if left alone and provided with sufficient funding. Unfortunately, such projects never end because the R&D managers either do not know when the project is over (poor understanding of the objectives) or do not want the project to end (exceeding objectives). In either event, periodic project review and readjustment action must be considered. The primary reason for periodic review is to reassess the risks based upon current strategic thinking and project performance. Souder has identified several types of project risks[7]:

- Technical failure
- Market failure

---

[7] Ibid.

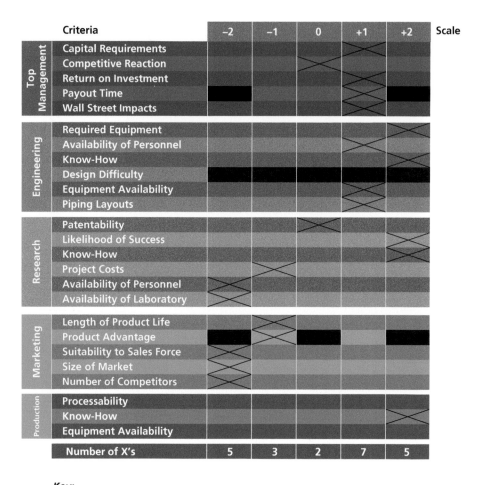

Key:

+2 = Excellent
+1 = Good
0 = Fair
−1 = Bad
−2 = Unacceptable

■ Not Applicable

⤬ Score for Project A

FIGURE 9-15 Illustration of a scaling model for one project

- Failure to perform
- Failure to finish on time
- Research failure
- Development failure
- Engineering failure
- Production failure
- User acceptance failure
- Unforeseen events

- Insurmountable technical obstacles
- Unexpected outcomes
- Inadequate know-how
- Legal/regulatory uncertainties

Project risks generally result in project selection readjustment. Typical readjustment actions might include:

- Replanning the project
- Readjusting the portfolio of projects
- Reallocating funds
- Rescheduling the project
- Backlogging the project
- Reprioritizing the projects
- Terminating the project
- Replacement with a backlogged project
- Replacement with a new project

9.18 PROJECT TERMINATION

Previously, we stated that R&D projects should be periodically reviewed so that readjustment actions can be taken. One technique for readjustment is simply to terminate the project. The following are the most common reasons and indications that termination is necessary:

- Final achievement of the objectives—This is obviously the best of all possible reasons.
- Poor initial planning and market prognosis—This could be caused by a loss of interest in the project by the marketing personnel or an overly optimistic initial strategy.
- A better alternative has been found—This could be caused by finding a new approach that has a higher likelihood of success.
- A change in the company interest and strategy—This could be caused by a loss of the market, major changes in the market, development of a new strategy, or simply a lack of commitment and enthusiasm of project personnel.
- Allocated time has been exceeded.
- Budget costs have been exceeded.
- Key people have left the organization—This could be caused by a major change in the technical difficulty of the project with the departure of key scientists who had the knowledge.
- Personal whims of management—This could be caused by the loss of interest by senior management.
- Problem too complex for the resources available—This could be caused by an optimistic initial view when, in fact, the project has insurmountable technological hurdles that did not appear until well into the project.

Executives normally employ one or more of the following methods to terminate the R&D projects:

- Orderly planned termination
- The hatchet (withdrawal of funds and cutting of personnel)
- Reassignment of people to higher priority projects
- Redirection of efforts toward different objectives or strategies
- Bury it or let it die on the vine

R&D project managers are highly motivated and hate to see projects terminate midstream. Executives must carefully assess the risks and morale effects of project termination.

9.19 TRACKING R&D PERFORMANCE

Throughout this book, we have stressed the fact that PM 2.0 uses more metrics than PM 1.0 for tracking and reporting project performance. Many of these traditional metrics may not provide an adequate representation of the status of an R&D project given the number of unknowns. Perhaps the hardest metrics to design are those that identify progress toward creativity. On R&D projects, creativity may not be known until the project is completed.

Sometimes, R&D projects are considered as failures because the desired results were not obtained. However, intellectual property could have been developed that leads to spinoffs for the creation of other products that are commercial successes. Because of this, it is often difficult to clearly define an R&D success and failure. Even if it is defined as a pure failure, but you detect early on that the results cannot be achieved, it can be seen as a success if you pull the plug early so as not to continue squandering money.

DISCUSSION QUESTIONS

The discussion questions are for classroom use to stimulate group thinking about PM 2.0. There are no right or wrong answers to most of the questions.

1. Can R&D projects be planned, scheduled, and controlled just like other projects?
2. Are all project managers qualified to manage R&D projects? If not, are there special skills that may be needed?
3. What is the role of R&D with regard to strategic planning?
4. What is the role of R&D with regard to the establishing of a portfolio of projects?
5. Are there different project management skills needed based upon whether the project is designed for offensive versus defensive R&D?
6. Who makes the final decision as to whether an R&D project must undergo an adjustment?
7. Who determines the priority for an R&D project?
8. What is the difference between a project life cycle and a product life cycle?
9. Can metrics be established to measure the success of an R&D project?
10. Can metrics be established to measure brainstorming and idea generation during R&D activities?

CHAPTER 10

PROBLEM SOLVING AND DECISION MAKING[1]

10.0 INTRODUCTION

The environment in which the project managers perform is changing significantly. Our projects are become more complex. The importance of time and cost has reached new heights in the minds of clients and stakeholders. Clients want to see the value in the projects they are funding. All of this is creating challenges for project managers in how they identify and resolve problems. To make matters more complex, project managers are now seen as managing part of a business when managing a project and are expected to make both project and business decisions.

Decisions are no longer a single-person endeavor as was often the case with PM 1.0. Project managers are expected to form problem-solving and decision-making teams. Most project managers have never been trained in problem solving, brainstorming, creative thinking techniques, and decision making. They rely on experience as the primary teacher. While that sounds like a reasonable approach, it can be devastating if project managers end up learning from their own mistakes rather than the mistakes of others. It is a shame that companies are unwilling to invest even small portions of their training budgets for these courses necessary for PM 2.0.

There are numerous books available on problem solving and decision making. Unfortunately, they look at the issues from a psychological perspective with applications not always relevant to project and program managers. Some books use the term problem analysis rather than problem solving. Problem analysis can be interpreted as simply looking at the problem and gathering the facts, but not necessarily developing alternative solutions for later decision making. In this chapter, problem solving is used throughout reflecting the identification of alternatives as well.

---

[1] Much of this chapter has been adapted from a work in progress, *Project-Based Problem Solving and Decision Making*, by H. Kerzner and C. Belack. Cark Belack can be reached at cbelack1@verizon.net and 1-978-266-3716.

10.1 UNDERSTANDING CONCEPTS

The world of PM 2.0, as described in Chapter 1, has imposed additional requirements on project managers. Without an understanding of both problem solving and decision making, project managers may struggle while managing projects.

Necessity for Problem Solving and Decision Making

We are forced to make decisions in our everyday life. We must decide what to eat, how to dress, where to go, when to go, and even who to socialize with. We may make 30 or more decisions a day. Some decisions, such as personal investment decisions, may be critical whereas other decisions may be just routine decisions. Most of these types of decisions we make by ourselves and usually have confidence in the fact that we made the right decision. And for some of the decisions, we can expend a great deal of time thinking through it.

But once we get to our place of employment, the decision-making process changes. We often have to involve many people in the process, some of whom we may never have met or worked with previously. The outcome of the decision can affect a multitude of people, many of which may be unhappy with the final result. The risks of a poor decision can lead to catastrophic consequences for the business. People who are unhappy with the decision and do not understanding it may view you now as an enemy rather than as a friend.

When we make personal decisions, we usually adopt a "let's live with it" attitude. If the decision is wrong, we may try to change the decision. But in a business environment, there may be a significant cost associated with changing a decision. Some business decisions are irreversible and can lead to a destruction rather than creation of business value.

But there is one thing we know for sure in a business environment: Anybody who always makes the right decision probably is not making enough decisions. Expecting to always make the right decision is wishful thinking. The wrong decision can be made regardless of the quantity and quality of information we have before us.

Problem solving and decision making go hand in hand. Decisions are made when we have issues or choices to make. In general, we must have a problem prior to making decisions. But there is a strong argument that decision making is needed and used as part of identifying the problem and developing alternatives. This is why most books discuss problem solving and decision making together.

Research Techniques in Basic Decision-Making Process

Human performance in the way we make decisions has been the subject of active research from several perspectives. There has been significant research in this area, and most results are part of four perspectives. The four basic perspectives are:

- The psychological perspective
- The cognitive perspective
- The normative perspective
- The problem-solving perspective

From a psychological perspective, it is necessary to examine individual decisions in the context of a set of needs, preferences an individual has, and values they seek. From a cognitive perspective, the decision-making process must be regarded as a continuous process integrated in the interaction with the environment. From a normative perspective, the analysis of individual decisions is concerned with the logic of decision making and rationality and the invariant choice to which it leads.

Yet, at another level that is perhaps more closely aligned with a project environment, it might be regarded as a problem-solving activity which is terminated when a satisfactory solution is found. Therefore, decision making is a reasoning or emotional process which can be rational or irrational and can be based on explicit assumptions or tacit assumptions. It is often impossible to separate discussions of decision making from problem solving. Yet both involve selecting among alternatives. The focus for most project managers is the problem-solving perspective, although in a project management environment, we could argue that all four perspectives somehow interact in the way we make decisions.

Facts about Problem Solving and Decision Making

There are several facts or generalities that we consider when discussing problem solving and decision making:

- Projects today are much more complex than before and so are the problems and decisions that must be made.
- Problem-solving techniques are used not only to solve problems but also to take advantage of opportunities.
- Today, we seem to be flooded with information to the point where we have information overload and cannot discern what information is actually needed or useful for solving problems.
- Lower to middle levels of management are often valuable resources to have when discussing the technical side of project problems. Senior management is valuable in the knowledge of how the problem (and its solution) relates to the overall business and the impact of the enterprise environmental factors.
- Problem solving today is a core competency, yet most companies provide very little training for their employees on problem solving and decision making.
- The project team may be composed of numerous subject matter experts, but the same people may not be creative and capable of thinking creatively for solving problems.
- The people that created the problem may not be capable of solving the problem they created.
- Few people seem to know the relationship between creativity and problem solving.

Information Overload

Today, with the additional project-based metrics that are available, there seems to be an abundance of information available to everyone. We all seem to suffer from information overload thanks to advances in information system technologies created with Web 2.0

technology. Our main problem is being able to discern what information is critical and what information should be discarded or stored in archives.

For simplicity sake, information can be broken down into primary and secondary information. Primary information is information that is readily available to us. This is information that we can directly access from our desktop or laptop. This is information that appears in a dashboard reporting system. Information that is company sensitive or considered as proprietary information may be password protected but still be accessible.

Secondary information is information that must be collected from someone else. Even with information overload, project managers generally do not have all of the information they need to solve a problem and make a timely decision. This is largely due to the complexity of our projects as well as the complexity of the problems that need to be resolved. We generally rely upon a problem-solving team to provide us with the secondary information. The secondary information is often more critical for decision making than the primary information. Many times the secondary information is controlled by the subject matter experts, and they must tell us what information is directly pertinent to this problem.

Collecting the information, whether primary or secondary information, can be time consuming. Information overload often forces us to spend a great deal of time searching through information when this time should be spent on problem solving. While dashboards show us performance, they do not identify what actions to take to correct an unfavorable situation. Additional information may be necessary.

Getting Access to Right Information

The project manager's challenge is not just getting the information, but getting the right information in a timely manner. Sometimes, the information that the project manager needs, especially secondary information, is retained by people who are not part of the problem-solving team. An example might be information related to politics, stakeholder relations management, economic conditions, cost of capital, and other enterprise environmental factors. This information may be retained by senior management or governance committee members.

Because timing is essential, project managers should have the right to talk directly with anyone they need to converse with to obtain the necessary information to solve a problem. Having to always go through the chain of command to access the information creates problems and wastes valuable time. Information is often seen as a source of power, which is one of the reasons why sometimes the chain of command must be followed in some companies.

Behind every door in a company is information of some sort. Project managers must be able to open those doors as needed. If project managers do not have access to those doors, then there are two options: follow the chain of command and hope that the information is not filtered by the time it gets to you or invite the person with this information to attend the problem-solving meeting. Where the person with the information resides in the organizational hierarchy determines their availability to attend the meeting. The higher up they reside, the less likely they will be able to attend your meeting in the near term. The project manager's accessibility to information is critical. Fortunately, one of the critical elements of PM 2.0 is active involvement in the project by the client and the governance committee.

Lack of Information

Even though we have information overload and access to secondary information sources, there is no guarantee we will have readily available all of the information we need. People that need to make decisions must accept the fact that they generally will not have all of the information they need on hand. This can happen at all levels of management, not just on projects. We must be willing to make the best possible decisions based upon the information we have at that time even if it is partial information. This will probably be the norm in most large PM 2.0 projects.

Too often, we rely on the chain of command for getting the information to help resolve a problem. If people believe that "possessing information is power," access to the needed information can be a problem, especially if they withhold some of the information. Because of the criticality of the constraints, time is not a luxury. Project managers must have the right or authority to access those people that possess the information. In PM 2.0, this is an absolute necessity. This assumes, of course, the project manager knows where the information resides. This is sometimes the greater challenge, especially if the needed information is nowhere to be found within the company. We must go outside the company to get the critical information.

Problem solving is most frequently based upon the best available information. Having all of the information needed to make a decision is wishful thinking.

Project versus Business Problem Solving and Decision Making

Project managers today believe that each project that they are managing is part of the business and therefore they are managing part of a business rather than just a project. As such, project managers are expected to make business decisions as well as project decisions. However, there is a difference between project decision making and business decision making:

- Project decision making focuses on meeting baselines, verification, and validation. Business decision making focuses on market share, profitability, customer satisfaction, and repeat business.
- Project decision making involves the project team whereas business decision making may include marketing, sales, and senior management.
- There are multiple tools that project managers use for project decision making, but the majority of the tools used for business decision making are mainly financial tools such as ROI, NPV, IRR, and cash flow.
- The focus of project decision making is project performance whereas the focus of business decision making is financial performance.
- The results of project decisions appear quickly whereas the results of business decisions may not appear for years.
- Most of the problems addressed as part of project decision making are to maintain the baselines whereas business problems focus on alignment or changes to the business strategy.

10.2 PROJECT ENVIRONMENT: ITS IMPACT ON PROBLEM SOLVING AND DECISION MAKING

In order to understand decision making in a project management environment, it is first necessary to understand how the project management environment differs from the traditional environment we are all used to. The project management environment is an interaction between people, tools, processes, and routine business work that must be accomplished for the survival of the firm and the completion of project work. Project management activities may be considered secondary to the ongoing business activities. Because of the high level of risks on many projects and the fact that some of the best resources are assigned to ongoing business activities, the decision-making process can lead to suboptimal or ineffective decisions.

There are other factors that make decision making quite complex in a project environment:

- The project manager may have limited or no authority at all to make the decisions even though they may have a serious impact on the project's outcome.
- The project manager does not have the authority to hire people to work on the project; they are assigned by functional managers after an often lengthy negotiation process.
- The people on the project team may not be able to make meaningful contributions to the decision-making process.
- The project manager may not have the authority to remove poor workers from the project team without assistance from the functional managers.
- The project manager may not have any responsibilities for wage and salary administration for the project team members; this is accomplished by the functional managers. Therefore, the project manager may not be able to exert penalty power if the workers make poor decisions.
- The team members are most likely working on other projects as well as your project, and you have no authority to force them to work on your project in a timely manner.

Because of the project manager's limited authority, some project managers simply identify alternatives and recommendations. These are then brought to the executive levels of management, the project sponsor, or the governance committee for the final decision to be made. However, some people argue that the project manager should have the authority to make those decisions that do not alter the deliverables of the project or require a change to the constraints and baselines.

Impact of Constraints on Project Problem Solving and Decision Making

The boundaries on most project decisions are the constraints that are imposed upon the project team at the initiation of the project. For decades, project managers focused primarily on the constraints of time, cost, and scope when making decisions. But in today's environment, we are undertaking more complex projects and many of these have significantly more than just three constraints. All of these constraints play havoc with the decision-making process. The time constraint probably has the greatest impact on decision making. Time is not a luxury. The decision may have to be made even though the project manager has only partial information. Making decisions with complete information is usually not a luxury that the project team will possess. And to make matters worse, we often have little knowledge of what the impact of the decisions will be.

Impact of Assumptions on Project Problem Solving and Decision Making

At the beginning of a project, it is imperative for the project team to know not only what limitations or constraints are imposed upon them but also what assumptions have been made. The assumptions are related to the enterprise environmental factors that surround the project. Usually the assumptions are listed in the project charter, but more often than not, it is just a partial list.

To make matters worse, assumptions can and will change over the life of most PM 2.0 projects. The longer the project, the more likely it is that the assumptions will change. Good project managers track the assumptions to see if they have changed.

The problem with having faulty assumptions is that they can lead to faulty conclusions, bad results, poor decision making, and unhappy customers. The best defense against poor assumptions is good preparation at project initiation, including the development of risk mitigation strategies and possibly a structured approach for how complex decisions will be made.

Assumptions have a major impact on the alternatives we select for the solution to a problem. Picking an alternative based upon faulty assumptions can have an undesirable impact on the project.

Understanding Project Environment

Understanding the project environment is essential. Some important factors to consider are:

- There are numerous constraints imposed upon the project.
- The constraints can change in relative importance over the life of the project and new constraints can appear.
- The assumptions made at the beginning of the project may no longer be valid as the project continues.
- The project manager may not know all of the constraints even though some of them are listed in the project charter.
- The project manager most likely does not have a command of technology in the area where the problem exists.
- The project team members may likewise not have a command of technology in the area where the problem exists.
- The project manager and the team are expected to make a decision in a timely manner even though they may not have complete information.

- The client and all of the associated stakeholders may not be in agreement with the final decision.
- Expecting to always make the right decision is wishful thinking.

Selecting Right Project Manager

Selecting the right project manager is essential. In general, the more complex the project, the more likely that problems will occur and, as expected, the problems will be quite complex. Unfortunately not all project managers possess problem-solving and decision-making skills, and it may be impossible at the initiation of a project to identify that these skills may be critical for this project. Some project managers are excellent in project execution but poor at problem solving. A history of experience on past projects seems to be the best way to select project managers who require these skills. These skills are not always easy to teach in a classroom.

Selecting a project manager with the right leadership style for certain projects is also essential. Some projects require that the project manager encourage the team to take risks, be creative, and be able to handle innovation. Unfortunately, the need for creativity and innovation may not be seen as a necessity at the onset of the project but may become a requirement when a problem occurs.

Some companies maintain a skills inventory database. At the end of each project, the project team is required to complete questionnaires that will be used to update the skills inventory database. The surveys include questions on creativity, problem solving, innovation, and decision making.

10.3 CONCEPTUAL PROBLEM-SOLVING AND DECISION-MAKING PROCESS

History has shown that project management performs significantly better when projects have a roadmap for all of the domain areas, namely initiation, planning, execution, monitoring and controlling, and closure. And, of course, included in each of the domain areas are problem-solving and decision-making activities, all involving some form of data gathering. Most project managers prefer some sort of structure beginning with data gathering. The roadmaps need not be based upon rigid policies and procedures but can be constructed using forms, guidelines, templates, and checklists. The latter provides the project team with significantly more flexibility when managing the project.

More companies today are developing their own processes for data gathering, problem solving, and decision making. The processes are supported by templates which undergo improvements from captured lessons learned and best practices. There is a large amount of similarity among companies in the steps they use for problem solving and decision making. Without these templates, data gathering, problem solving, and decision making become ad hoc rather than structured processes.

Determining the Steps

Rational thinkers prefer an analytical approach to data gathering, problem solving, and decision making using sequential steps. There are several steps a company can choose from when setting up an approach:

- Recognizing the problem
- Understanding the problem
- Gathering the data
- Understanding the environmental impacts
- Understanding the assumptions
- Understanding the constraints
- Understanding the boundaries on the problem and the solution
- Convening the problem-solving team if not already done
- Generating alternatives
- Redefining the assumptions and constraints
- Evaluating the trade-offs
- Evaluating the impact of the solution
- Selecting the best option
- Getting approval of the option
- Implementing the alternative
- Monitoring and controlling the solution

Companies currently perform many of these steps, but all of the steps may not be clearly defined as part of the company's approach. Also, many of the steps may be done in parallel rather than sequentially.

Given the complexity of PM 2.0 projects, the following six steps and the order of the steps may be more appropriate as a starting point:

1. **Identifying and Understanding the Problem:** Identify the real problem, ensuring you are not just addressing a symptom of the problem; once properly identified, understand the depth of the problem and its implications for the project.
2. **Gathering the Data:** Using various techniques, gather all of the data associated with the problem, doing your best to weed out nonrelated data.
3. **Analyzing the Data:** Based on the type of data and available requisite expertise, evaluate the data to understand the true root cause(s) of the problem.
4. **Developing Alternative Solutions:** Develop multiple approaches that might be used to resolve the problem.
5. **Decision Making—Selecting the Best Solution:** Using appropriate tools and methods and avoiding associated psychological pitfalls, decide which of the identified solutions is the optimal one for the project to implement.
6. **Evaluating the Decision and Taking Corrective Action:** Once the decision has begun to be implemented, monitor the outcome(s) and take remedial action if needed.

It should be understood that, while the above steps are shown in a linear fashion, many of these steps are performed iteratively. For example, let us assume that the project manager and his or her team have identified what they believe to be the problem. They then begin to gather the data and start its analysis. It is quite possible that the data will point to a different problem than that which has been identified. If that is the case, additional data may have to be gathered and further analyzed.

As another example, at some point after a decision that selects the "best" alternative is accomplished and the team starts to implement the decision, it may be determined that the selected solution is not working as intended. Consequently the team may decide they need to take a different approach to the problem—one that may have been a previously identified but not selected approach or a novel approach not previously discussed by the team. Additionally, the process may be truncated prior to running its full course. The team may get through the first three steps and realize that no solution to the problem exists or there may only be one solution to the problem, eliminating the need for decision making.

As we go through each of the steps, we will describe various methods, tools, and techniques that have been developed for each of the steps, focusing primarily, however, on solution development and decision making.

10.4 IDENTIFYING AND UNDERSTANDING A PROBLEM

To understand problem solving, we must first understand what is meant by a problem. A problem is a deviation between an actual and desired situation. It is an obstacle, impediment, difficulty, or challenge or any situation that invites resolution, the resolution of which is recognized as a solution or contribution toward a known purpose or goal. The problem could be to add something that is currently absent but desired, to remove something that is potentially bad, or to correct something that is not performing as expected. Therefore, problems can be formulated in a positive or negative manner. Problems are formulated in a positive manner if the problem is to determine how to take advantage of an opportunity.

We tend to identify alternatives as being good or bad choices. If the decision maker has all of the alternatives labeled as good or bad, then the job of the decision maker or project manager would be easy. Unfortunately, a problem implies that there exists doubt or uncertainty or else a problem would not exist. This uncertainty can happen on all projects and therefore makes it difficult to classify all alternatives as only good or bad.

The time it takes to resolve a problem is often a measure of the complexity of the problem. Adding more people to the problem-solving session cannot always reduce the time to resolve the issue. In some cases, we end up creating additional problems for ourselves. Some problems must be resolved whereas other problems may be able to be delayed and then resolved sometime in the future. Some problems are "good" problems because the objective is to take advantage of an opportunity.

Real Problems versus Personality Problems

We generally believe that most problems are real and need to be resolved. But that is not always the case. Some problems are created based upon the personalities of the individuals.

Some people create problems unnecessarily as long as they can somehow benefit, perhaps by being the only person capable of solving the problem. Some examples are:

- Resolution of the problem will get you more power.
- Resolution of the problem will get you more authority.
- Resolution of the problem will diminish the power and authority of others.
- You are the only one with the capability to resolve the problem and it will improve your image and reputation.
- You will be regarded as a creative thinker.
- It will look good on your resume.
- It will look good during performance reviews.
- It will guarantee you employment.

It is important to discover first of all whether or not the problem is real and if a simple solution exists for it. Many years ago, a department manager was afraid that downsizing would take place and that he would lose his position as a department manager. To protect himself, he gave the workers assigned to project teams conflicting instructions knowing that problems would occur and that several projects might suffer. This resulted in rework and created problems for several projects. The department manager then called a problem-solving meeting with those project managers who were falling behind on their projects. The department manager stated in the meeting that almost all of his employees were poor workers who needed constant supervision and that the problem would be resolved by the department manager. He would provide these poor workers with much closer supervision. The department manager provided alternatives to the fictitious problem and stated that he would have the problems resolved within a few months.

The department manager believed that his position was now secure. But the project managers were not fooled. The project managers found out the truth and eventually the department manager was fired for what he did. The project managers discovered that this was not a real problem that needed to be resolved using problem-solving and decision-making techniques.

Not All Problems Can Be Solved

Problems imply that some alternatives exist. Problems that have no alternatives are called open problems. Not all problems can be solved or should be solved. As an example, in R&D and new product development, it may take as many as 50–60 ideas to generate one commercially new product. The cost could be prohibitive to evaluate this many alternatives. Another common problem is in software development where "gold plating" the project with additional features that are unnecessary could have a serious impact on the end date.

Some problems cannot be resolved without a breakthrough in technology. Companies that do not possess the technical skill or financial resources to undertake these breakthrough projects leave them as open projects. The same can also hold true for projects involving company image, reputation, and goodwill.

Finally, there are those projects that require compliance to government regulations. For these projects, which are almost always very costly, all of the alternatives are often considered poor choices. When forced to comply, we select the best of the worst. But more often than not, we leave them as open problems until the very last minute hoping the problem will be forgotten or disappear.

Complexity of Problems

Not all problems have the same degree of complexity. Problem complexity determines whether we should address the problem or leave it as an open problem. Some factors used to identify problem complexity are:

- Relative magnitude of the problem
- The cost to resolve the problem
- The availability of qualified resources to be part of the problem-solving team
- How well the problem is understood
- The amount of information available to solve the problem
- Whether we have partial or complete information on the problem
- The amount of remaining work on the project that may be impacted by the solution to the problem
- How the client will view the solution to the problem
- How the stakeholders will view the solution to the problem
- The impact that the solution (or failure of the solution) can have on the project team members' careers
- Whether the team will be motivated to find a solution to the problem
- Whether viable alternatives can be found for a solution

Technique for Problem Identification

It is impossible to solve a problem or take advantage of an opportunity without first being able to identify that a problem exists. Most people know that a problem exists when the project fails to meet the baselines or when dashboard metrics display unfavorable performance. Waiting for the problem to surface may limit the time available for finding the best solution. Techniques available to the project team for problem identification include:

- Selecting the right metrics and key performance indicators
- Using templates and checklists that provide a series of questions that can be asked to identify present or future problems
- Effective use of team meetings
- Using walk-the-halls project management
- Monitoring the enterprise environmental factors for critical changes
- Working with people involved in customer relations management programs and customer value management programs
- Performing project health checks
- Listening to complaints
- Establishing a project suggestion box
- Performing risk management and establishing risk triggers
- Reading literature related to your industry

Individual Problem Solving Conducted in Secret

Companies encourage all project team members to bring forth all problems quickly. The quicker the problem is exposed, the more time is available for finding a solution, the more alternatives are usually available, and the greater the number of resources that can assist in the solution. Unfortunately, some people simply do not want to openly identify the problem with the hope that they can resolve it by themselves before anyone finds out about it. This is true for people who may have been involved in creating the problem. Reasons for this include:

- Damage to one's reputation and image
- Damage to one's career
- Loss of employment
- Able to solve the problem using their own ideas rather than the ideas of others
- Dislike asking for help from others
- Distrust of the solution that others might choose
- Fear of antagonism from colleagues and team members
- Preference to working alone rather than in a team

In such cases, people try to solve the problem by themselves, in secret, before anyone finds out about the problem. In reality, the problem is often hard to hide.

Team Problem Solving Conducted in Secret

Sometimes the entire problem-solving team is in collusion in hiding the problem. Unfortunately, problem-solving sessions clearly identify that a problem exists and this alone could make it difficult to hide a problem. It is easier for one person to try to solve a problem secretly than an entire team.

Based upon the severity of the problem, the actual problem could be withheld from the client, the stakeholders, and even your own management, although the latter is certainly not a good idea. Sometimes people are not informed even after a solution is found and implemented. There are several reasons for wanting the problem to be resolved quietly:

- The client and/or the stakeholders and/or the governance committee may overreact to the problem and dictate the solution.
- The client and/or key financial stakeholders may overreact to the problem and remove financial support.
- The client may cancel the project.
- Problem resolution requires the discussion of proprietary or classified information.
- Open identification of the problem may cause people to be fired.
- Open identification of the problem may cause damage to your company's image and reputation.
- Open identification of the problem can result in potential lawsuits.
- The cause of the problem is unknown.
- The problem can be resolved quickly without any impact on the competing constraints and the deliverables.

10.5 GATHERING PROBLEM-RELATED DATA

Data-gathering techniques are a necessity in all domain areas of the *PMBOK® Guide* and, as such, can be considered a crossover skill that is needed by everyone involved in the project. It may be difficult or even impossible to make the right decision without having all of the necessary information available. The larger and more complex the project, the greater the need for effective data-gathering skills. Unfortunately, the organizational process assets and enterprise project management methodologies that project teams use do not necessarily provide all of the information needed to manage a project.

As an example of the use of data gathering, as part of the scope management area of knowledge in the *PMBOK® Guide*, data-gathering techniques are used to collect the requirements, which is the process of documenting the stakeholders' needs to meet the project's objectives. Without effective data gathering, it may be impossible to meet the customer's and stakeholders' expectations.

When managing a project, the project manager must not expect that all of the information needed will be provided by the organization process assets. Although not often explicitly defined, it is the job of the project manager to gather all of the necessary information whether it is done individually or through team members.

Reason for Data Gathering

In addition to collecting requirements, data-gathering techniques are used to:

- Decide what decisions to make
- Decide what the impact of the decisions might be
- Decide what action items are necessary
- Determine the root cause of a problem
- Determine the causes for both favorable and unfavorable variances from a plan
- Determine the number and grade level of the resources that are needed
- Determine the risks that can occur and how they will be managed
- Select vendors
- Negotiate contracts

While everyone strongly supports the use of metrics in project management, it should be understood that metrics are simply measurements against some standard or baseline. Metrics may identify part of the problem, but additional information must be discovered. As an example, having an unfavorable cost variance may be seen as a problem, but it is the reason for the cost variance that is the true problem.

Data-Gathering Techniques

There are numerous techniques available for data gathering. The selection of the technique is based upon the information being sought out, the timing of the information, who will provide the information, the criticality of the information, and the type of decisions that the information must support. Each technique comes with strengths and weaknesses. Some data-gathering techniques can be done quickly. For the most part, data-gathering techniques are time consuming. A partial list of techniques might be:

- Root-cause analysis
- Panels of experts
- Facilitated work groups
- Questionnaires
- Surveys
- Interviews
- Observations and measurements
- Use of prototypes
- Diagramming techniques such as cause-and-effect diagrams
- Critical reviews of performance data
- Case study analysis

Using just one technique may not suffice. It may be necessary to use several techniques in order to capture all of the required data.

Setting Limits on Problem Solving and Decision Making

Problem solving and decision making can go on for a long period of time. Limitations must be established early on. Some of the limits are:

- How much time is available to address the issue
- How much money the project is willing to commit to resolve the problem
- How many resources can be assigned to resolve the problem
- Whether the assigned resources will have the required skills
- Which facilities can be used for testing or other activities
- The importance of the project to the company
- The importance of the project to the client
- The importance of the project to the stakeholders
- The criticality of the project and the problem

It is not uncommon for the limitations to be identified in the problem statement or even in the agenda. When people understand the limitations early on, decisions are usually made in a timely manner.

Identifying Boundary Conditions

Limitations or boundary conditions must also be established for the solution to the problem, and the limitations can impact the alternatives that will be selected. We know for a fact that, when solving a problem and identifying alternatives, we do not have an infinite sum of money to implement a solution or as much time as we want. We consider these as constraints, but they are also boundary conditions. Boundary conditions can be established by the client, stakeholders, and eventual users of the deliverables. A partial list of boundary conditions might be:

- Staying within the project's constraints on time, cost, quality, and scope or within all of the competing constraints
- Without increasing the risks on the remaining work on the project
- Without altering the scope of the remaining work on the project

- Without altering the company's normal flow of work
- Without gold-plating the solution
- Without including unnecessary functionality
- Knowing that only a limited number of additional resources are available for solving the problem and implementing the solution
- Without violating regulatory agency requirements such as those established by OSHA and EPA
- Without driving up the selling price of the product beyond what the customers will pay for it

Determining Who Should Attend Problem-Solving Meeting

Problems are not resolved in a vacuum. Meetings are needed and the hard part is to determine who should attend. If people are involved in the problem or the problem is unrelated to the work they do, then having them attend these meetings may be a waste of their time. This holds true for some of the team members as well. As an example, if the problem is with procurement, then it may not be necessary for the drafting personnel to be in attendance.

For simplicity sake, we shall consider just two types of meetings: problem solving and decision making. The purpose of the problem-solving meeting is to obtain a clear understanding of the problem, collect the necessary data, and develop a list of workable alternatives accompanies by recommendations. More than one meeting will probably be required.

Sending out an agenda is important. The agenda should include a problem statement which clearly explains why the meeting is being called. If people know about the problem in advance, they will have a chance to think about the problem and bring the necessary information, thus reducing some of the time needed for data gathering. It is also possible that the information gathered will identify that the real problem is quite different from what was considered to be the problem at first.

It is essential that subject matter experts familiar with the problem be in attendance. These subject matter experts may not be part of the original project team but may be brought in just to resolve this problem. The subject matter experts may also be contractors hired in to assist with the problem.

The people brought in for the identification of the problem and data gathering usually remain for the development of the alternatives. But there are situations where additional people may participate just for the consideration of alternatives.

Determining Who Should Attend Decision-Making Meeting

The decision-making meeting is different from the problem-solving meeting. In general, all of the participants that were involved in the problem-solving meeting will most likely be in attendance in the decision-making meeting, but there may be a significant number of other participants. Project team members should have the ability to resolve problems, but not all of the team members have the authority to make decisions for their functional units. It is normally a good idea at the initiation of the project for the project manager to determine which team members possess this authority and which do not.

Team members that do not possess decision-making authority will still be allowed to attend the decision-making sessions but may need to be accompanied by their respective functional managers when decisions are required and voting takes place.

Stakeholder attendance is virtually mandatory at the decision-making meetings. The people making the decisions must have the authority to commit resources to the solution of the problem. The commitment could involve additional funding or the assignment of subject matter experts and higher pay grade employees.

Project managers are responsible for the implementation of the solution. Therefore, the project manager must have the authority to obtain the resources needed for a timely solution to the problem.

Creating Framework for Meeting

For the problem-solving meeting, it is important to create a mental framework of the problem beforehand, including what should be accomplished in the meeting and the limitations. Not all of the people who will attend the problem-solving meeting will be familiar with the problem. Some may have just a cursory understanding of the problem and others may not have known that the problem even existed prior to the meeting.

The mental framework should include all of the information known thus far about the problem. Additional information will most likely be forthcoming in the meeting. If possible, the framework should be included in the invitation for the meeting and/or the agenda. Informing people about the problem prior to the meeting will get them to think about it and possibly even perform some preliminary research prior to the meeting. When people understand the framework prior to the meeting, they usually come to the meeting better prepared and may even recommend to the organizer of the meeting other people who should be invited.

Understanding How People React in Meetings

Team meetings that involve problem solving and decision making often get people to act in an irrational manner, especially if the outcome of the meeting can have a negative impact on them personally. This is particularly true for people who are closely identified with the cause of the problem. You may also be inviting people you have never worked with previously and you have no idea how they will react to the problem or the solution. Attitudes that must be closely watched in these meetings are:

The Aggressor: Criticizes everyone as being part of the problem even if the cause is just one person.

The Devil's Advocate: Always argues that there are other causes of the problem and refuses to become a believer in the real cause unless threatened.

The Dominator: Tries to take over the meeting and professes to know everything about the problem and what the solution should be. This is seen as a chance for glory.

The Recognition Seeker: Always argues in favor of his or her impression of the problem and also his or her solution to the problem.

The Withdrawer: May be afraid of criticism and does not want to be identified as being part of the problem.

There are other attitudes that can appear in meetings, but these are the ones that seem to occur most frequently in project management.

Working with Participants during Meetings

If problems must be identified and alternatives must be found, then working with people in the traditional manner may not work. This is particularly true if you have people in the meeting with the attitudes discussed previously. In these types of meetings, there will be bickering and conflicts. The project manager or meeting leader must create an environment that will lead to a successful outcome. Given the makeup of the participants and the severity of the problem, there are things that the project manager can say or does that would make it easier for the people to participate. Some of the expressions that the project manager can use are:

- Is there a chance that this might be the real cause or is there a chance that this might work?
- Have we tried anything like this before?
- Do we know any other time where this has happened?
- Do we know of any other companies that have had similar problems?
- Your idea has a lot of merit!
- Your idea is great but we may have to make a small change.
- What you said will really help us.
- Are we saying that. . . .
- Let me say in my own words what I think you just said.
- Let's see if we can put this in perspective.
- Your idea and my idea are close together.
- Aren't we saying the same thing?
- Let's see if we are in agreement.
- Let's see how the rest of the team feels about this.
- Who hasn't given us their opinion yet?
- Are we prepared to make a decision? Or is there additional information we need?
- Should we keep our options open?

It should be obvious in most of these expressions that the project manager is trying to solicit feedback.

Based upon the severity of the problem to be resolved, the chairperson of the meeting may be someone other than the project manager. The chairperson may be someone specifically trained in facilitation skills. The project manager would attend the meeting but not function as the chairperson.

Leadership Techniques during Meetings

There are several techniques that the leader or the project manager can adopt to encourage the best possible outcome from the meetings. These include:

- Encouraging people to speak
- Asking probing questions
- Avoiding questions that may be counterproductive
- Keeping emotions under control
- Soliciting feedback
- Providing constructive feedback rather than personal criticism
- Understanding the people on the team and their needs and interests
- Understanding the legal implications of all of the alternatives and decisions
- Resisting attempts at gamesmanship

Handling Problem-Solving and Decision-Making Conflicts

Conflicts and conflict resolution are a way of life in project management. Some conflicts may have a higher intensity level and are therefore more difficult to resolve. Not all conflicts are bad. People often argue in favor of their own opinion, and if they continue to bring forth additional facts that may be important, then the conflict may be allowed to continue. These types of conflicts are often seen as "constructive" conflicts.

During problem-solving and decision-making sessions, it is often desirable to invite people that you know are opponents to attend these sessions. These people generally bring forth a significant amount of data to support their position and quite often the additional information causes a change to the alternatives selected.

Not everyone will be in agreement with the source of the problem, the alternatives, or the decision. There is a chance that some of the conflicts may linger on long after the final decision is made. Expecting everyone to agree with the final decision is wishful thinking. In an event, the project manager or meeting chairperson must have reasonable knowledge in conflict management. If a great deal of conflict is expected at the meetings, it may be desirable to have a professional facilitator act as the chairperson rather than the project manager.

Continuous Solutions versus Enhancement Project Solutions

Not all problems need to be resolved immediately. Some problems may be needed to be resolved immediately whereas other problems can be clustered together and resolved later using enhancement projects. An example might be a software project that is designed to support inventory management at a company's manufacturing plants. After the project was well underway, the plants wanted changes to be made to the software. Resolving all of the issues as they came up was deemed unrealistic by the project manager because the launch date of the original project would be delayed continuously. In this case, the project manager was able to establish an enhancement project that would address many of these issues as a separate project that would be completed sometime after the original project was completed. Not all problems can be pushed into the future for resolution.

Not only can there be a schedule slippage with continuous problem solving, but there may be a lack of qualified resources. Most companies do not have excess resources sitting by idly waiting for an assignment. Unavailable resources can further elongate the schedule.

Problem Solving versus Scope Creep

Too often, project managers believe that most problems must be resolved using scope changes. The result is scope creep. Scope creep is the continuous enhancement of the project's requirements as the project's deliverables are being developed. Scope creep is viewed as the growth in the project's scope and is often rationalized as a necessity to solve all of the problems that come up when managing a project.

Scope creep is a natural occurrence, and we must accept the fact that it will happen. Scope creep can lead to beneficial results. But, on the other hand, we can argue that scope creep is not just allowing the scope to change but also an indication of how well we prevent unwanted changes from occurring. In this regard, project managers should ask themselves:

- Do we need scope changes to resolve this problem?
- Is the customer always right with regard to their view of the solution?
- Have we acted as the devil's advocate to see what would happen without a scope change?
- Will a solution that requires a scope change lead to other scope changes?

Problem Solving and Decision Making during Crisis Projects

Crisis problem solving and decision making are significantly different from ordinary project problem solving and decision making. Crisis projects are those projects that either have already or may result in the loss of life. Examples include loss of life resulting from product tampering and the use of faulty equipment. Some of the major differences in problem solving and decision making when human life is involved are:

- Time is an extremely critical constraint rather than just an ordinary constraint.
- Life-cycle phases are measured in hours or days rather than weeks or months.
- The chairperson of the problem-solving and decision-making teams may be the project sponsor or someone from senior management levels rather than the project manager.
- Sufficient time may not exist for a complete understanding of the root cause of the problem.
- The problem-solving and decision-making meetings may include representation from all of the stakeholders, even those that were mainly observers in the past.
- The news media may be actively interested in what you are doing to solve the problem and extreme care must be taken on how to communicate with them.
- Communications with the media will normally be handled by someone from senior management rather than the project manager.
- The alternative selected to resolve the problem must be able to be implemented as quickly as possible.
- Future risks and the potential of future lawsuits may be an important factor in selecting among alternatives.

10.6 ANALYZING DATA

Most projects do not get into trouble overnight. By selecting the right metrics for a project, there are early warning signs that a problem is about to materialize. It is much easier to solve problems when they are small than when they become large.

At the beginning of a project, we perform risk management. Part of the risk management process is to identify potential problems that may occur on the project and to establish metrics and risk triggers that provide an early warning sign that a problem might occur. To assist us in doing this, we have lessons learned files, best practices and metrics libraries, and diaries on previous projects.

Establishing core metrics early on in the project for many of the problems that might occur is certainly a good idea. However, it may not be practical. The more metrics you have, the more costly it becomes to track, measure, and report these metrics. But having some metrics is certainly better than having no metrics at all.

Having a metric that identifies a cost overrun or schedule slippage is nice to have but it does not identify the cause of the problem. But if you accompany this metrics with another metric that identifies the number of assigned resources or the quality (pay grade) of the assigned resources, you may have a better understanding of the cause of the problem or where to begin looking.

Questions to Ask

Effective data gathering requires an understanding of what questions to ask. While it is true that the questions will be predicated upon the type of problem, typical questions might be:

- Are there any other resources or subject matter experts that can help us with this problem?
- How many problems do we have?
- Are there hidden problems that are below the surface?
- What is the extent of the problem?
- Is the problem getting worse, getting better, or stable?
- Did this problem exist previously?
- Can the problem be quantified?
- Can we determine the severity of the problem?
- What physical evidence exists to identify the problem?
- Who identified the problem?
- To whom was it first reported?
- Is there an action plan to collect additional information?
- Do we have the right team members addressing this problem?

10.7 DEVELOPING ALTERNATIVE SOLUTIONS

A major part of problem solving and ultimately decision making involves the identification and analysis of a finite set of alternatives described in terms of some evaluative criteria. These criteria may be benefit or cost in nature or the criteria could simply be the adherence to the cost, schedule, and scope baselines of the project. Then the problem might be to rank these alternatives in terms of how attractive they are to the decision maker(s) when all the criteria are considered simultaneously. Another goal might be to just find the best alternative or to determine the relative total priority of each alternative.

The number of alternatives is often limited by the constraints imposed upon the project. For example, if the actual schedule is exceeding the baseline schedule, then the project manager may have five alternatives: overtime, performing some work in parallel rather than in series, adding more resources to the project, outsourcing some of the work to a lower cost supplier, or reducing the scope of the project. Each alternative will be accompanied by advantages and disadvantages. If the goal is to lower the costs, then there may be only one viable alternative, namely reducing the scope.

Because of the complexity of projects, the project manager cannot be expected to determine all of the alternatives in a vacuum. The team should be involved in the identification and prioritization of the alternatives. If the team does not have the expertise to determine the alternatives, then the project manager may need additional support from the subject matter experts in the functional areas. There are also situations where the stakeholders or external contractors may be able to provide the necessary information for determining alternatives.

Variables to Consider during Alternative Analyses

There are several variables that must be considered when identifying and selecting alternatives. The variables are usually project specific and based upon the size, nature, and complexity of the problem. However, we can identify a core list of variables that usually apply to the identification and evaluation of most alternatives:

- **Cost:** There is a cost associated with each alternative. This includes not only the cost of implementing the alternative but also the financial impact on the remaining work on the project.
- **Schedule:** Implementing an alternative takes time. If the implementation time is too long or cannot be done in parallel with the other project work, then there may be a significant impact on the end date of the project.
- **Quality:** Care must be taken that the speed to resolve a problem does not result in a degradation of quality in the project's deliverables
- **Resources:** Implementing a solution requires resources. The problem is that the people needed with the necessary skills may not be available.
- **Feasibility:** Some alternatives may seem plausible on paper but may be unfeasible when needed to be implemented. Feasibility or complexity of the alternative must be considered. Otherwise, you could make the problem worse.
- **Risks:** Some alternatives expose the company to increased risks. These may be future risks (or even opportunities) that will appear well after the project is completed.

Understanding Features That Are Part of Alternatives

Previously, we discussed some of the variables that must be considered when looking at alternatives. Now, we look a little deeper into the features that are part of the alternatives. Many times there are several features that can be included in each of the alternatives. Part of understanding the boundary conditions is to know the importance of each feature. The features can be classified as:

- **Must Have:** Any alternative that does not include this feature should be discarded.
- **Should Have:** These are features that in most situations should be included in the alternatives that are being considered. Failure to consider these could result in a degradation of performance. Some of these features may be omitted if including them results in unfavorable consequences when trying to satisfy the competing constraints.
- **Might Have:** These are usually add-ons to enhance performance but not necessarily part of the project's requirements. These are nice-to-have items but not a necessity when deciding upon a final solution. Might-have features are often characterized as bells and whistles that are part of gold-plating efforts.

Developing Hybrid Alternatives

After looking at the variables and evaluating all of the alternatives, the conclusion may be that none of the alternatives are acceptable. In this case, the project manager may be forced to select the "best of the worst." As an example, consider a utility which must comply with standards imposed by the U.S. Environment Protection Agency. In this case, the company is quite unhappy with all of the alternatives, but, by law, the problem must be resolved and one of the alternatives must be selected.

It is possible that, after evaluating the alternatives, the best approach might just be a combination of alternatives. This is referred to as hybrid alternatives. Alternative A might be a high risk but a low cost of implementation. Alternative B might be a low risk but a high cost of implementation. By combining alternatives A and B, we may be able to come up with a hybrid alternative with an acceptable cost and risk factor.

Trade-Offs

Trade-offs are decision-making exercises that most frequently result in changing or sacrificing one part of the project to gain in another part of the project. Hybrid alternatives are examples of trade-offs. Trade-offs are most often made on the competing constraints of the project. For example, if the problem is to maintain the quality of the deliverables, we may need to provide more funding, allow for more time, or both. We also use trade-off analysis techniques when we have several seemingly good alternatives and try to get the best features of each alternative condensed into a single alternative. This involves trade-offs needed to combine alternatives rather than a trade-off between alternatives.

While many of the people who sit in the problem-solving and decision-making sessions are experts in their field and come up with realistic alternatives to solve a problem, they often lack the ability to understand the trade-offs that are necessary and the impact. For example, the solution to meeting the client's quality requirements might be adding more time into the schedule for some additional work, but the cost could be prohibitive. Also, there could be a financial impact on other suppliers that are tied into our schedules.

The project team is usually best qualified to evaluate trade-offs on alternatives even though people outside of the team were used to identify the alternatives. People outside of the project team may be brought in to address a specific problem and may not see the entire picture and the impact of their recommendations.

Common Mistakes When Developing Alternatives

The alternative selection process is subject to errors as is any other problem-solving process. The errors could be by chance or intentional. Examples of errors that plague the selection process are:

- The time and cost estimates for a particular alternative are grossly underestimated to make this alternative look highly attractive to the decision makers. The people who provided the estimates may have done this intentionally because of personal gains or recognition if this alternative is selected. This is being done at the expense of the project.
- Support for an alternative is done with overoptimism to the point where the true implementation risks are hidden. This could be the most expensive alternative and the client will then be required to fund lucrative scope changes.
- Support for a good alternative is done with underoptimism in hopes that the client with pick one of the more costly alternatives.

10.8 PROBLEM-SOLVING TOOLS AND TECHNIQUES[2]

There are problem-solving tools and techniques available. Sometimes, more than one approach is selected to see if the results are the same. Each tool comes with advantages and disadvantages.

Root-Cause Analysis

Root-cause analysis (RCA) is a class of problem-solving methods aimed at identifying the root causes of problems or events. The practice of RCA is predicated on the belief that problems are best solved by attempting to address, correct, or eliminate root causes, as opposed to merely addressing the immediately obvious symptoms. By directing corrective measures at root causes, it is more probable that problem recurrence will be prevented. However, it is recognized that complete prevention of recurrence by one

---

[2] Many of the techniques discussed here have been taken from Wikipedia, the free encyclopedia.

corrective action is not always possible. Conversely, there may be several effective measures (methods) that address the root cause of a problem. Thus, RCA is often considered to be an iterative process and is frequently viewed as a tool of continuous improvement.

RCA is typically used as a reactive method of identifying event(s) causes, revealing problems, and solving them. Analysis is done *after* an event has occurred and the problem is visible. Insights in RCA may make it useful as a proactive method. In that event, RCA can be used to *forecast* or predict probable events even *before* they occur. While one follows the other, RCA is a completely separate process.

Root-cause analysis is not a single, sharply defined methodology; there are many different tools, processes, and philosophies for performing RCA analysis. However, several very broadly defined approaches or "schools" can be identified by their basic approach or field of origin: safety based, production based, process based, failure based, and systems based.

General Principles of RCA

There are general principles of RCA. They include:

- The primary aim of RCA is to identify the root cause(s) of a problem in order to create effective corrective actions that will prevent that problem from ever recurring, otherwise addressing the problem with virtual certainty of success. ("Success" is defined as the near-certain prevention of recurrence.)
- To be effective, RCA must be performed systematically, usually as part of an investigation, with conclusions and root causes identified backed up by documented evidence. Usually a project team effort is required.
- Although there may be more than one root cause for an event or a problem, the difficult part is demonstrating the persistence and sustaining the effort required to develop them.
- The purpose of identifying all solutions to a problem is to prevent recurrence at lowest cost in the simplest way. If there are alternatives that are equally effective, then the simplest or lowest cost approach is preferred.
- Root causes identified depend on the way in which the problem or event is defined. Effective problem statements and event descriptions (as failures, for example) are helpful, or even required.
- To be effective, the analysis should establish a sequence of events or timeline to understand the relationships between contributory (causal) factors, root cause(s), and the defined problem or event to prevent it in the future.
- RCA can help to transform a reactive culture (that reacts to problems) into a forward-looking culture that solves problems before they occur or escalate. More importantly, it reduces the frequency of problems occurring over time within the environment where the RCA process is used.
- RCA is a threat to many cultures and environments. Threats to cultures often meet with resistance. There may be other forms of management support required to achieve RCA effectiveness and success. For example, a "nonpunitory" policy toward problem identifiers may be required.

Corrective Actions Using RCA

RCA forms the most critical part of successful corrective action because it directs the corrective action at the true root cause of the problem. The root cause is secondary to the goal of prevention, but without knowing the root cause, we cannot determine what an effective corrective action for the defined problem will be. Steps to consider include:

- Define the problem or describe the event factually.
- Gather data and evidence and then classify that along a timeline of events to the final failure or crisis.
- Ask "why" and identify the causes associated with each step in the sequence toward the defined problem or event.
- Classify causes into causal factors that relate to an event in the sequence and root causes that, if applied, can be agreed to have interrupted that step of the sequence chain.
- If there are multiple root causes, which is often the case, reveal those clearly for later optimum selection.
- Identify corrective action(s) that will with certainty prevent recurrence of the problem or event.
- Identify solutions that, if effective, prevent recurrence with reasonable certainty with consensus agreement of the group, are within your control, meet your goals and objectives, and do not cause or introduce other new, unforeseen problems.
- Implement the recommended root-cause correction(s).
- Ensure effectiveness by observing the implemented recommendation solutions.
- Other methodologies for problem solving and problem avoidance may be useful.

RCA Techniques

RCA tools and techniques include:

- Barrier analysis—a technique often used in process industries. It is based on tracing energy flows, with a focus on barriers to those flows, to identify how and why the barriers did not prevent the energy flows from causing harm.
- Bayesian inference.
- Causal factor tree analysis—A technique based on displaying causal factors in a tree structure such that cause–effect dependencies are clearly identified.
- Change analysis—An investigation technique often used for problems or accidents. It is based on comparing a situation that does not exhibit the problem to one that does in order to identify the changes or differences that might explain why the problem occurred.
- Current reality tree (CRT)—A method developed by Eliahu M. Goldratt in his theory of constraints that guides an investigator to identify and relate all root causes using a cause–effect tree whose elements are bound by rules of logic (categories of legitimate reservation). The CRT begins with a brief list of the undesirable things we see around us and then guides us towards one or more root causes. This method is particularly powerful when the system is complex,

there is no obvious link between the observed undesirable things, and a deep understanding of the root cause(s) is desired.
- Failure mode and effects analysis.
- Fault tree analysis.
- Five Whys: Ask why, why, why, why, and why over again until exhausted.
- Ishikawa diagram, also known as the fishbone diagram or cause-and-effect diagram. The Ishikawa diagram is used by project managers for conducting RCA as well as for resolving quality and risk problems.
- Pareto analysis "80/20 rule."

Other tools and techniques exist in addition to what is listed here.

Brainstorming

Throughout the life of any project, the team will be tested on their ability to find the best possible solution to a problem within the imposed limitations and boundaries. This could occur in the planning phase of the project where we must come up with the best possible approach for a plan or it could happen in any later phases where problems arise and the best solution must be found. These are situations where brainstorming techniques may not be appropriate. Most people seem to have heard about brainstorming but very few have been part of brainstorming teams.

Although brainstorming has become a popular group technique, when applied in a traditional group setting, researchers have not found evidence of its effectiveness for enhancing either the quantity or quality of ideas generated. Although traditional brainstorming does not increase the productivity of groups (as measured by the number of ideas generated), it may still provide benefits, such as boosting morale, enhancing work enjoyment, and improving teamwork. Thus, numerous attempts have been made to improve brainstorming or use more effective variations of the basic technique.

Although we normally discuss brainstorming as a means of identifying alternative solutions to a problem, brainstorming can also be used for root-cause identification of the problem.

Rules for Brainstorming

There are four basic rules in brainstorming. These rules are intended to stimulate idea generation and increase overall creativity of the group while minimizing the inhibitions people may have about working in groups.

Focus on Quantity: This rule focuses on the maximization of possible ideas, both good and bad. The assumption made is that the greater the number of ideas, the greater the chance of finding the optimal solution to a problem.

Withhold Criticism: In brainstorming, criticism of ideas creates conflict and wastes valuable time needed to generate the maximum number of ideas. When people see ideas being criticized, they tend to withhold their own ideas to avoid being criticized. Criticism should take place, but after the brainstorming session is completed. Typical brainstorming sessions last about an hour or less.

Welcome Unusual Ideas: All ideas should be encouraged, whether good or bad. People must be encouraged to think "out of the box," and this may generate new perspectives and a new way of thinking. Sometimes, what appears as a radical solution initially may be the best possible solution in the end.

Combine and Improve Ideas: The best possible solution may be a combination of ideas. New ideas should be encouraged from the combination of ideas already presented.

Critical Steps in Brainstorming

There are several critical steps that must occur for brainstorming to be successful. The following has been adapted from Wikipedia, the free encyclopedia:

- **Set the Problem:** Before a brainstorming session begins, it is critical to define the problem. The problem must be clear, not too big, and captured in a specific question. If the problem is too big, the facilitator should break it into smaller components, each with its own question.
- **Create a Background Memo:** The background memo is the invitation and informational letter for the participants, containing the session name, problem, time, date, and place. The problem is described in the form of a question, and some example ideas are given. The memo is sent to the participants well in advance, so that they can think about the problem beforehand.
- **Select Participants:** The facilitator composes the brainstorming panel, consisting of the participants and an idea collector. A group of 10 or fewer members is generally more productive. Many variations are possible, but the following composition is suggested:
  - Several core members of the project who have proved themselves
  - Several guests from outside the project, with affinity to the problem
  - One idea collector who records the suggested ideas
- **Session Conduct:** The facilitator leads the brainstorming session and ensures that ground rules are followed. The steps in a typical session are:
  1. A warm-up session, to expose novice participants to the criticism-free environment. A simple problem is brainstormed; for example, *can we minimize the number of reports on this project? Or what can be improved in the way we do verification and validation?*
  2. The facilitator presents the problem and gives a further explanation if needed.
  3. The facilitator asks the brainstorming group for their ideas.
  4. If no ideas are forthcoming, the facilitator suggests a lead to encourage creativity.
  5. All participants present their ideas, and the idea collector records them.
  6. To ensure clarity, participants may elaborate on their ideas.
  7. When time is up, the facilitator organizes the ideas based on the topic goal and encourages discussion.
  8. Ideas are categorized.
  9. The whole list is reviewed to ensure that everyone understands the ideas.

10. Duplicate ideas and obviously infeasible solutions are removed.
11. The facilitator thanks all participants and gives each a token of appreciation.

Conducting Brainstorming Session: Process

The process of conducting a brainstorming session includes the following:

- Participants who have ideas but were unable to present them are encouraged to write down the ideas and present them later.
- The idea collector should number the ideas, so that the chairperson can use the number to encourage an idea generation goal; for example: *We have 14 ideas now, let's get it to 20!*
- The idea collector should repeat the idea in the words he or she has written verbatim to confirm that it expresses the meaning intended by the originator.
- When many participants are having ideas, the one with the most associated idea should have priority. This is to encourage elaboration on previous ideas.
- During a brainstorming session, functional managers and other superiors may be discouraged from attending, since it may inhibit and reduce the effect of the four basic rules, especially the generation of unusual ideas.

Conducting Brainstorming Session: Evaluation

Brainstorming is not just about generating ideas for others to evaluate and select. Usually the group itself will, in its final stage, evaluate the ideas and select one as the solution to the problem proposed to the group:

- The solution should not require resources or skills the members of the group do not have or cannot acquire.
- If acquiring additional resources or skills is necessary, that needs to be the first part of the solution.
- There must be a way to measure progress and success. This may require establishing new metrics.
- The steps to carry out the solution must be clear to all and amenable to being assigned to the members so that each will have an important role.
- There must be a common decision-making process to enable a coordinated effort to proceed and to reassign tasks as the project unfolds.
- There should be evaluations at milestones to decide whether the group is on track toward a final solution.
- There should be incentives to participation so that participants maintain their efforts.

Brainstorming Sessions: Nominal Group Technique

There are several variations in the way brainstorming sessions are conducted:

- The nominal group technique is a type of brainstorming that encourages all participants to have an equal say in the process. It is also used to generate a ranked list of ideas.

- Participants are asked to write their ideas anonymously. Then the moderator collects the ideas and each is voted on by the group. The vote can be as simple as a show of hands in favor of a given idea. This process is called distillation.
- After distillation, the top ranked ideas may be sent back to the group or to subgroups for further brainstorming. For example, one group may work on the color required in a product. Another group may work on the size, and so forth. Each group will come back to the whole group for ranking the listed ideas. Sometimes ideas that were previously dropped may be brought forward again once the group has reevaluated the ideas.
- It is important that the facilitator be trained in this process before attempting to facilitate this technique. The group should be primed and encouraged to embrace the process. Like all team efforts, it may take a few practice sessions to train the team in the method before tackling the important ideas.

Group-Passing Technique

- Each person in a circular group writes down one idea and then passes the piece of paper in a clockwise direction to the next person, who adds some thoughts. This continues until everybody gets his or her original piece of paper back. By this time, it is likely that the group will have extensively elaborated on each idea.
- The group may also create an "Idea Book" and post a distribution list or routing slip to the front of the book. On the first page is a description of the problem. The first person to receive the book lists his or her ideas and then routes the book to the next person on the distribution list. The second person can log new ideas or add to the ideas of the previous person. This continues until the distribution list is exhausted. A follow-up "read out" meeting is then held to discuss the ideas logged in the book. This technique takes longer, but it allows individuals time to think deeply about the problem.

Team Idea-Mapping Method

- This method of brainstorming works by the method of association. It may improve collaboration and increase the quantity of ideas and is designed so that all attendees participate and no ideas are rejected.
- The process begins with a well-defined topic. Each participant brainstorms individually; then all the ideas are merged onto one large idea map. During this consolidation phase, participants may discover a common understanding of the issues as they share the meanings behind their ideas. During this sharing, new ideas may arise by the association, and they are added to the map as well. Once all the ideas are captured, the group can prioritize and/or take action.

Electronic Brainstorming

- Electronic brainstorming is a computerized version of the manual brainstorming technique. It is typically supported by an electronic meeting system (EMS), but simpler forms can also be done via email and may be browser based or use peer-to-peer software.

- With an electronic meeting system, participants share a list of ideas over the Internet. Ideas are entered independently. Contributions become immediately visible to all and are typically anonymous to encourage openness and reduce personal prejudice. Modern EMSs also support asynchronous brainstorming sessions over extended periods of time as well as typical follow-up activities in the creative-problem-solving process such as categorization of ideas, elimination of duplicates, and assessment and discussion of prioritized or controversial ideas.

- Electronic brainstorming eliminates many of the problems of standard brainstorming, production blocking, and evaluation apprehension. An additional advantage of this method is that all ideas can be archived electronically in their original form and then retrieved later for further thought and discussion. Electronic brainstorming also enables much larger groups to brainstorm on a topic than would normally be productive in a traditional brainstorming session.

- Some Web-based brainstorming techniques allow contributors to post their comments anonymously through the use of avatars. This technique also allows users to log on over an extended time period, typically one or two weeks, to allow participants some "soak time" before posting their ideas and feedback. This technique has been used particularly in the field of new product development but can be applied in any number of areas where collecting and evaluating ideas would be useful.

Directed Brainstorming

- Directed brainstorming, which is similar to a technique called brain writing, is a variation of electronic brainstorming (described previously). It can be done manually or with computers. Directed brainstorming works when the solution space (that is, the criteria for evaluating a good idea) is known prior to the session. If known, those criteria can be used to intentionally constrain the ideation process.

- In directed brainstorming, each participant is given one sheet of paper (or electronic form) and told the brainstorming question. They are asked to produce one response and stop, and then all of the papers (or forms) are randomly swapped among the participants. The participants are asked to look at the idea they received and to create a new idea that improves on that idea based on the initial criteria. The forms are then swapped again and respondents are asked to improve upon the ideas, and the process is repeated for three or more rounds.

- In the laboratory, directed brainstorming has been found to almost triple the productivity of groups over electronic brainstorming.

Individual Brainstorming

- "Individual brainstorming" is the use of brainstorming on a solitary basis. It typically includes such techniques as free writing, free speaking, word association, and drawing a mind map, which is a visual note-taking technique in which people diagram their thoughts. Individual brainstorming is a useful method in creative writing and has been shown to be superior to traditional group brainstorming under many circumstances.

Question Brainstorming

- This process involves brainstorming the *questions*, rather than trying to come up with immediate answers and short term solutions. This technique stimulates creativity and promotes everyone's participation because no one has to come up with answers. The answers to the questions provide the framework for constructing future action plans. Once the list of questions is set, it may be necessary to prioritize them to reach the best solution in an orderly way. Another problem for brainstorming can be to find the best evaluation methods for a problem.

Wikipedia provides an excellent list of references for brainstorming.

10.9 CREATIVITY AND INNOVATION

You are placed in charge of a project that is quite complex and perhaps even a high risk. Some sort of technical breakthrough may be required for the project to be regarded as a success. You team keeps coming up with problems that need to be resolved. How does a project manager know if the assigned resources have creative abilities? This is an essential skill for problem solving.

Not all people are creative even if they are at the top of their pay grade. People can do the same repetitive task for so long that they are considered subject matter experts. They can rise to the top of their pay grade based upon experience and years of service. But that alone does not mean that they have creativity skills. Most people think that they are creative when, in fact, they are not. Companies also do not often provide their workers training in creative thinking.

In a project environment, creativity is the ability to use one's imagination to come up with new and original ideas or things to meet requirements and/or solve problems. People are assigned to project teams based upon experience. It is impossible for the project manager, and sometimes even the functional managers, to know whether these people have the creativity skills needed to solve problems that can arise during a project. Unless you have worked with these people previously, it is difficult to know if people have imagination, inspiration, ingenuity, inventiveness, vision, and resourcefulness, all being common characteristics of creativity.

In project management, creativity is the ability to think up ideas to produce something new through imaginative skills, whether a new solution to a problem or a new method or device. Innovation is the ability to solve the problem by converting the idea into reality, whether it is a product, service, or any form of deliverable for the client. Innovation goes beyond creative thinking.

Creativity and innovation do not necessarily go hand in hand. Any problem-solving team can come up with creative solutions that cannot be implemented. Any engineering team can design a product (or a modification to a product) that manufacturing cannot build.

Although most people seem to believe that innovation is directly related to discoveries made by the R&D group, innovation also involves contributions made by all business functions of an organization (sales and marketing, finance, operations, etc.) toward the solution to a problem. Simply stated, innovation, as part of problem solving, is a team effort.

Creativity, Innovation, and Value

Innovation is more than simply turning an idea into reality. It is a process that creates value. Clients are paying for something of value. Whatever solution is arrived at must be recognized by the client as possessing value. The best of all possibilities is when the real value can be somehow shared between the client's needs and your company's strategy. The final alternative selected might increase or decrease the value of the end deliverables, as seen by the client, but there must always be some value in the solution selected.

Some solutions to a problem may necessitate a reduction in value compared to the original requirements of the project. This is referred to as negative innovation. In such cases, innovation for a solution that reduces value can have a negative or destructive effect upon the team. People could see negative innovation as damage to their reputation and career.

If the innovation risks are too great, the project team may recommend some form of open innovation. Open innovation is a partnership with those outside your company by sharing the risks and rewards of the outcome. Many companies have creative ideas for solving problems but lack the innovative talent to implement a solution. Partnerships and joint ventures may be the final solution.

Negative Innovation

Sometimes, we start out projects with the best of intentions and later discover that some problem has occurred that could result in the cancellation of the project. Rather than cancel the project outright, the solution might be to downsize the project and readjust our innovation attempts. Factors that can lead to a readjustment in innovation include:

- The market for the deliverable has shrunk.
- The deliverable will be overpriced and demand will not be there.
- The technical breakthrough cannot be achieved in a timely manner.
- There is a loss of faith and enthusiasm by the team and they no longer believe this solution is workable.
- There is a possible loss of interest by top management and the client.
- Insurmountable technical obstacles exist.
- There is a significant decrease in the likelihood of success.

If these factors exist, then it is entirely possible that another alternative must be selected in order to salvage the project. As long as the client is willing to accept a possible reduction in final value, the project may be allowed to continue.

Types of Innovative Solutions

There are several types of innovation. Each type comes with advantages and disadvantages. The three most common are:

Product/Quality Improvements and Cost Reduction Efforts: This type of innovation may be able to be accomplished quickly and with the existing resources in the company. The intent is to solve a problem and add incremental value to the end result.

Radical Breakthrough in Technology: This type of innovation has risks. You may not be able to determine when the breakthrough will be made and the accompanying cost. Even if the breakthrough can be made, there is no guarantee that the client will receive added value from this solution. If the breakthrough cannot be made, the client may still be happy with the partial solution. This type of innovation may require the skills of only one or two people.

Totally Complex System or Platform: This is the solution with the greatest risk. If the complex system cannot be developed, then the project will probably be considered as a total loss. A large number of highly talented resources are needed for this form of innovation.

Problem-Solving and Decision-Making Attributes That Are Difficult to Teach

There are courses on problem solving and decision making. Unfortunately, there are some attributes that excellent problem solvers and decision makers possess that cannot be taught easily in courses. The three most common attributes are:

Instinct: The innate behavior that is the inherent inclination toward a particular type of behavior. It is an inborn pattern of behavior that is characteristic of a response to specific stimuli such as the desire to solve complex problems. It is often used with such words as a natural inclination, a natural or acquired tendency, aptitude, talent, knack, gift, capacity, genius, faculty, intuition, feeling, impulse, gut feeling, or sixth sense. People that have this instinct thrive on problem solving. The more complex the problem, the greater their voluntary involvement.

Common Sense: Sound, practical judgment derived from experience rather than specialized knowledge. People that possess common sense tend to make decisions based upon the situation and the facts rather than based upon their technical knowledge. However, they may rely upon past experiences.

Guesswork: The process of making guesses, or finding the conclusions arrived at by guessing. Guesswork is often used with other words such as conjecture, deduction, presumption, speculation, estimation, reasoning, approximation, and estimation. Guesswork is required when an estimate must be made and little or no information is available. Based upon the problem, guesswork may be the very first step when initiating problem-solving or decision-making sessions. Sometimes, guesswork is required simply to understand the problem.

Creative Roadblocks

There are also blockages to being creative. These include:

- Not understanding the problem well enough and attacking the wrong problem
- Making assessments and decisions too quickly
- Taking the first idea that is acceptable to the team
- Having a team that considers you as an outsider
- Having a team that refuses to support any of your ideas
- Having a team that has no faith in your ability to be part of the team

These roadblocks do not necessarily apply only to problem solving that involves innovation. Also, the solution to some problems simply does not require innovation.

10.10 DECISION MAKING: SELECTING BEST SOLUTION

We must now decide which of the alternatives best resolves the problem. More often than not, decision making has some degree of structure to it. Decision making involves the following:

- Objectives must first be established.
- Objectives must be classified and placed in order of importance.
- Alternative actions must be developed.
- The alternatives must be evaluated against all the objectives.
- The alternative that is able to achieve all the objectives is the tentative decision.
- The tentative decision is evaluated for more possible consequences.
- The decisive actions are taken, and additional actions are taken to prevent any adverse consequences from becoming problems and starting both systems (problem analysis and decision making) all over again.

The decision-making activities are often more time consuming and costlier to perform than the problem-solving activities. This is largely due to the number of alternatives that can be identified and the methods used to evaluate and prioritize them. Having a significant number of reasonable alternatives may seem nice, but being unable to arrive at a decision on which one to actually adopt can be troublesome.

There are several types of decision-making styles that people can use. There are also numerous tools that can assist in the decision-making processes.

Understanding How Decisions Are Made

People must understand prior to attending both the problem-solving and the decision-making meetings how the decisions will be made. There are several options available and the approach taken to agree on the problem can be different from the decision on which alternative will be adopted. Options include:

Majority or Consensus: All participants in the meeting are allowed to vote. The criterion might be a simple majority or another number such as a 75% majority.

Qualified Majority or Consensus: If a majority is not reached, then the project manager, the client, or another designated individual will make the final decision.

Project Manager Directed: The project manager makes the decision and informs the team which alternative he or she selected. This approach is most effective on crisis projects.

Client Directed: The team identifies the alternatives, makes a recommendation, and presents the data to the client. The client then makes the final decision and informs the team. The client may have the right to not select from the team's alternatives and to develop their own solution to the problem.

Routine Decision Making

Some decisions are easy to make whereas others require teams of experts. The tools and techniques used are dependent upon the type of decision. As an example, let us consider three types of decisions:

- Routine decisions
- Adaptive decisions
- Innovative decisions

Routine decisions are often handled solely by the project manager. Routine decisions may involve simply signing purchase orders, selecting which vendors to work with, and deciding whether or not to authorize overtime. Usually routine decisions are based upon company policies and procedures.

While routine decisions seem relatively easy to make, the number of routine decisions can be troublesome. Too many routine decisions can become time robbers and prevent the project manager from effectively managing the project. If the decisions are routine in nature, then many of the decisions may be able to be delegated to members of the project team.

Adaptive Decision Making

Adaptive decision making may require some degree of intuition. The problem is usually well understood and the project team may be able to make the decision without outside support or sophisticated tools and techniques. Adaptive decision making is the most common form of decision making used on projects. Examples are:

- Determining the number of tests that should appear in the test matrix
- Determining when an activity should begin or end
- Determining how late an activity can start without delaying downstream work
- Determining how late raw materials can be ordered
- Determining whether the work should take place on regular shift or overtime
- Determining whether a risk management plan is necessary, and if it is necessary, how much detail should appear in the place
- Determining how often testing should take place to validate compliance to quality requirements

- Determining the resource skill set needed, assuming there are choices
- Determining the best way to present both good news and bad news to the stakeholders
- Determining ways to correct unfavorable cost and schedule variances
- Determining the leadership style to be used to motivate certain team members
- Determining how to best reward superior performance by team members.

Innovative Decision Making

Innovation is generally regarded as a new way of doing something. The new way of doing something should be substantially different from the way it was done before rather than a small incremental change such as with continuous improvement activities. The ultimate goal of innovation is to create hopefully long-lasting additional value for the company, the users, and the deliverable itself. Innovation can be viewed as the conversion of an idea into cash or a cash equivalent.

Innovative decision making is most often used on projects involving R&D, new product development, and significant product enhancements. These decisions involve subject matter experts that may not be part of the project team and may require the use of more advanced decision-making tools and techniques. These decisions may require a radical departure from the project's original objectives. Not all project managers are capable of managing projects involving innovation by themselves.

While the goal of successful innovation is to add value, the effect can be negative or even destructive if it results in poor team morale, an unfavorable cultural change, or a radical departure from existing ways of doing work. The failure of an innovation project can lead to demoralizing the organization and causing talented people to be risk avoiders in the future rather than risk takers.

Pressured Decision Making

Time is a critical constraint on projects, and this can have a serious impact on the time necessary to understand the problem and find a solution. As an example, assume that a critical test fails and the client says that they will be meeting with you the day after the failure to discuss how you will correct the problem. They are expecting alternatives and a recommendation.

Typically, you might need a week or longer to meet with your team and diagnose the situation. However, given the circumstances, you may have to make a decision, right or wrong, based upon the time available. This is high-pressured decision making. Given sufficient time, we can all analyze or even overanalyze a problem and come up with a list of viable alternatives.

High-pressured decision making can also be part of adaptive and innovative decision making as well. Being pressured to make a decision can have favorable results if it forces the decision makers to look at only those attributes that are critical to the problem. But more often than not, high-pressured decision making leads to suboptimal results.

Given that these situations will happen, you must expect that you will not always have complete or perfect information in order to make a decision. Most decision-making teams must deal with partial information.

Decision-Making Meetings

While some decisions that are routine or adaptive in nature can be handled during regular team meetings, in general problem resolution team meetings should be set up as separate meetings. The attendance at a problem resolution meeting can be quite different from the attendance at regular team meetings. Stakeholders and clients may be required to attend problem resolution team meetings since they are the people most likely affected by the decision. Functional managers and subject matter experts may also be invited to attend. Outside consultants with critical expertise may also participate.

There is a wide variety of decision-making tools and techniques that project teams can use for decision making during these meetings. The selection of the best tool or technique to use can be based upon the complexity of the problem, the risks associated with the decision, the cost to make the decision and the impact if the decision is wrong, to whom the decision is important, the time available to make a decision, the impact on the project's objectives, the number of people on the project team, the relative importance to the customer or stakeholders, and the availability of supporting data.

Generally speaking, more than one meeting may be required. The purpose of the first meeting may be to just understand the problem and gather the facts. The problem-solving team may then require additional time to think through the problem and identify alternatives. It is highly unlikely that a decision will be made at the first team meeting.

Decision-Making Stages

There are several models available for how project teams make decisions. A typical four-stage model might be:

- **Familiarization Stage:** This is where the team meets to understand the problem and the decision(s) that must be made.
- **Options Identification Stage:** This is where the team performs brainstorming and lists possible alternatives for a solution.
- **Option Selection Stage:** This is where the team decides upon the best option. The team selecting the preferred option may have a different makeup than the team developing the list of options.
- **Justification Stage:** This is where the team rationalizes that they made the right decision and possibly evaluates the results.

Decision-Making Steps

When a project team is faced with a difficult decision, there are several steps one can take to ensure the best possible solution will be decided. There are several models that can be used. Many of the steps are similar to the steps used in setting up the problem-solving process. Steps that are common to many of these models include:

- Send out an agenda identifying the purpose of the decision-making meeting and the expected outcome, including the expected timing for the decision.
- Ask the team to come to the meeting with alternative solutions, if possible.

- Perform a RCA of the problem to make sure everyone understands why a decision is necessary and the impact if a decision is not made.
- Have the team brainstorm alternatives, including the option for combining alternatives.
- List the pros and cons for each alternative.
- Select the best option.
- Explain the decision to those involved and affected.
- Implement the decision.
- Measure the impact of the decision and capture/document best practices and lessons learned.

Advantages of Group Decision Making

As stated before, there are situations such as routine decisions where the ultimate decision is made by just the project manager. Groups are not needed in this situation. But more often than not, the problems that appear on projects require group thinking.

There are several advantages to group decision making. These include:

- Groups provide better decisions than individuals.
- Group discussions lead to a better understanding of the problem.
- Group discussions lead to a better understanding of the solution.
- Groups make better judgment calls on the selection of alternatives.
- Groups tend to accept more risks in problem solving than do individuals.
- Clients appear less likely to question the decision of the group compared to the decision of an individual.
- People are more willing to accept the final decision if they participated in the decision-making process.

Disadvantages of Group Decision Making

There are several disadvantages to group decision making. These include:

- The discussions can be dominated by the personality of one person regardless of whether or not that person is regarded as a subject matter expert.
- Groups may accept too much risk knowing that a failure would be blamed equally among all of the members of the group.
- There is pressure to accept the decision of the group even though you know that other decisions might be better.
- Too much time may be spent arriving at a consensus.
- Groups tend to overthink problems and solutions.
- It may be impossible to get the proper people released from other duties so they can attend the meeting.
- Finding a common meeting time that satisfies all parties may be difficult.
- If external people are involved, the costs associated with traveling could become quite large, especially if more than one decision-making meeting is needed.

Rational versus Intuitive Thinking

Decision making requires thinking. There are four forms of thinking that we will discuss. Rational thinking, often referred to as analytical thinking, refers to logical or reasoning being involved in the thought processes required for problem solving and decision making. It refers to providing reasons or the rationale behind thoughts or ideas. It adds an element of calculation and planning to a stream of thoughts rather than basing them on emotion or personal opinion. It is a kind of objective process of thinking and an analytic approach to any problem. Rational thinking is based on reasons or facts and is hence much more calculating and realistic. All people are capable of thinking rationally, but people will tend to cloud this ability because of emotions, prejudices, and possibly a fear of making a decision.

Rational thinkers believe that problems are easier to solve if they are broken down into well-defined sequential steps. They maximize the use of forms, guidelines, templates, and checklists. Rational or analytical thinking is efficient when there is sufficient time, a relatively static condition, and a clear differentiation between the observer and the observed. It is best suited for dealing with complexities and works best where there are established criteria for the analysis (for example, rules of law).

Intuitive thinking has contrasting qualities: It is unfocused and nonlinear, contains "no time," sees many things at once, views the big picture, contains perspective, is heart centered and oriented in space and time, and tends to the real or concrete. Intuition comes into its own where analytical thinking is inadequate: under time pressure, where conditions are dynamic, and where the differentiation between observer and observed is unclear. Intuitive thinking is a necessity when working on projects. It works best when seeking the "best" option in favor of the "workable" and when the project team is prepared to act on feelings or hunches where either explanations are not required or there is no time for them. Intuition is experience translated by expertise to produce rapid action.

Divergent versus Convergent Thinking

Divergent thinking is a thought process or method used to generate creative ideas by exploring many possible solutions by expanding the problem and looking at the big picture. It is often used in conjunction with convergent thinking, which is a narrow-focused detailed picture that follows a particular set of logical steps to arrive at one solution, which in some cases is a "correct" solution. Divergent thinking typically occurs in a spontaneous, free-flowing manner such that many ideas are generated. Many possible solutions are explored in a short amount of time, and unexpected connections are drawn. After the process of divergent thinking has been completed, ideas and information are organized and structured using convergent thinking. Divergent thinking moves away in diverging directions so as to involve a variety of aspects and sometimes lead to novel ideas and solutions. It is usually associated with creativity. Idea generation techniques such as brainstorming and out-of-the-box thinking are used in which an idea is followed in several directions to lead to one or more new ideas, which in turn lead to still more ideas.

In contrast to divergent thinking, which is creative, open-ended thinking aimed at generating fresh views and novel solutions, convergent thinking aims at bringing together information focused on solving a special problem (especially solving problems that have a single correct solution).

Polarity Management

There are times when it appears that you have an either/or choice to make to help resolve a problem, but it turns out that such a choice will, in fact, be a false choice. While selecting one of the two choices may resolve the issue in the short run, over time the dilemma pops up again. It may be that what you are facing is not an "either/or" problem to be resolved but rather a "both/and" polarity that needs to be properly managed.

Polarity management is required when you have two polar opposites that, at least to some extent, are both part of the solution. Assume you have been assigned to manage a large, complex project and are trying to figure out what the best management style is to use with your project team. On the one hand, you need to maintain control over the project, so a hierarchical approach might seem the appropriate way to go. However, you do not want to stifle the initiative of the team members you have selected for their expertise and experience on previous related projects, and you want to allow them to take the lead when required by the course of the project.

A number of such polarities exist within the business and project domains, including clarity/ambiguity, rigidity/flexibility, team/individual, etc. Try thinking of others and you will quickly see how frequently these polarities occur. A polarity to be managed can be identified through its (1) persistence over time and (2) the existence of interdependent poles. If a "problem" exists over a period of time and seems irresolvable and you can identify the simultaneous need for its polar opposite, chances are it is an issue of polarity management.

Fear of Decision Making: Mental Roadblocks

Not everyone wants to make decisions or is capable of making them. Some people would prefer to have others make all decisions, especially critical decisions. Reasons for this behavior might include:

- A previous history of making the wrong decisions
- Emotional fear of making the wrong decision
- Fear of the associated risks
- Lack of conviction in one's own beliefs
- Having high levels of anxiety
- Unable to cope with the politics of decision making
- Unfamiliar with the facts surrounding the problem and not willing to learn
- Unfamiliar with members of the team
- Possessing poor coping skills
- Lack of motivation
- Lack of perspective

- Being brought into the discussion well after the discussion began
- Unable to work under high levels of stress and pressure
- Afraid of working with unions that are involved in the problem
- Afraid of working with certain stakeholders involved in the problem
- Afraid of contributing for fear of being ridiculed
- Afraid of exposing one's inadequacies
- Afraid of damaging your career and/or reputation

These roadblocks are often categorized into five areas:

- Emotional blockages
- Cultural blockages
- Perceptual blockages
- Intellectual blockages
- Expressive blockages

Danger of Hasty Decisions

The project's constraints often place the project manager in a position of wanting to make hasty decisions. Making hasty decisions is sometimes a necessity, but more often than not, the results can be detrimental. Hasty decisions can lead to:

- Additional problems surfacing later in the project
- Rework that leads to cost overruns and schedule slippages
- Excessive overtime
- Customers and stakeholders that will lose faith in your ability to manage the project correctly
- Lack of faith in the problem-solving and decision-making process
- Manpower curves with peaks and valleys rather than a smoothed-out manpower curve
- Greater hands-on involvement by the governance committee
- More meetings
- An increase in reporting requirements
- Deliverables that are rejected by the client

Simply stated, speed in decision making is a risky business.

Decision-Making Styles

Not all decisions are easy to make. Sometimes you must make a decision whether you are ready or not and when you have partial rather than complete information available to you. Also, the decision to do nothing differently may be the best decision under certain circumstances. If the team believes that they can live with the problem at hand, then the team may wait and see if the problem gets worse before making a decision.

Every project manager has their own approach to decision making and this may vary from project to project. The style selected is based upon the definition of the problem and the type of decision that must be made. Although some approaches work well, there are approaches that often do more harm than good.

Textbooks on decision making provide several different styles. The five styles most common for project managers are:

- The autocratic decision maker
- The fearful decision maker
- The circular decision maker
- The democratic decision maker
- The self-serving decision maker

Autocratic Decision Maker

The autocratic decision maker usually trusts nobody on the team and dictates the decision even though the risks are great and very little time was consumed discussing the problem. Team members often are fearful of presenting alternatives and recommendations because they may be ridiculed by the project manager who believes that his or her decision is the only one. Team members may not contribute ideas even when asked.

The autocratic style can work if the project manager is regarded as an expert in the area in which the decision must be made. But, in general, project managers today seem to possess more of an understanding of technology than a command of technology. As such, using the autocratic style when you have limited knowledge about the technology of the problem and the solution can lead to a rapid decision but often a decision that is not the optimal choice.

Most of the time, autocratic decision makers feel better making a decision by themselves without any input from others. They make a decision on the spot based upon a feeling in their gut. This is often a hit-or-miss approach.

Fearful Decision Maker

While the autocratic decision maker thrives on making the decision, right or wrong, and in a timely manner, the fearful decision maker is afraid of making the wrong decision. This is often referred to as the "ostrich" approach to making a decision. In this case, the project manager will bury his or her head in the sand and hope that the problem will disappear or that people will forget about the problem. The project manager also hopes that, by waiting, a miracle solution will appear by itself such that a decision may not have to be made at all.

Sometimes the fearful decision maker adopts a procrastination attitude, which is waiting for enough (or at least a minimum amount of) information so that a decision can be made. This does not necessarily mean avoiding a decision. The fearful decision maker knows that a decision must be made, eventually.

The fearful decision maker is afraid that making the wrong decision could have a serious impact upon his or her reputation and career. The team may not be invited to provide alternatives and recommendations because that would indicate that a problem exists and that a decision must be made. Information on the problem may even be withheld from senior management, at least temporarily.

The project manager may try to get others to make the decision. The project manager may prefer to have someone act as the moderator of the decision-making group and, if a decision must be made, the project manager will always argue that it is a group decision rather than a personal decision. The project manager will avoid, if possible, taking personal accountability and responsibility for the decision.

As stated previously, time is a constraint on projects, not a luxury. Taking a wait-and-see approach to making a decision can lose precious time where the problem could have been easily resolved. Also, the longer we wait to make a decision, the fewer the options are.

Circular Decision Maker

The circular decision maker is similar to the fearful decision maker. The project manager not only wants to make the decision but also wants to make the perfect decision. Numerous team meetings are held to discuss the same problem. Each team meeting seems to discuss the problem and possible solutions from a different perspective. The team members are given action items that keep them scurrying about looking for additional information to support the perfect decision.

The circular decision maker is willing to make a decision but sacrifices a great deal of time looking for the perfect decision that everyone will agree to. The decision maker is willing to violate the time constraints on a project to accomplish this. The decision maker may also believe that the problem may disappear if they think about it long enough.

The project manager can adopt the circular decision-making style even if he or she is an expert in the area in which the problem exists. The project manager needs reenforcement from the team, and possibly superiors, that the best decision was made. In the eyes of the project manager, the decision may be deemed more important than the outcome of the project.

Democratic Decision Maker

The democratic decision maker allows the team members to participate in the final decision. Voting by the group membership is critical and may even be mandatory. The company may even have a structured approach for this using guidelines or templates. This can happen even if the project manager is the expert in the area where the problem exists and even if the project manager has the authority to make the decision by himself or herself.

Democratic decision making can create long-term problems. Team members may feel that they should be involved in all future decisions as well, even those where they may have limited knowledge about the problems. Asking team members to take an early vote on the solution to a problem can lead to apprehension if the team members are uncomfortable with making a decision based upon incomplete information. Waiting too long to make the decision can limit the options available and frustrate the project team because of the time that was wasted overthinking the problem and the solution.

Democratic decision making is a strong motivational tool if used properly. As an example, if the project manager believes that he or she already knows that decision that should be made, asking the team for their opinion and giving credit to a team member for coming up with the same idea is a good approach. This encourages people to participate in decision making and makes them believe that they will be given credit for their contributions.

Self-Serving Decision Maker

Everyone sooner or later is placed in a position where they must decide what is more important when making a decision: their individual values or the organizational values. This situation often forces people to make decisions in favor of either themselves or the organization. A compromise might be impossible.

These types of self-serving conflicts can permeate all levels of management. Executives may make decisions in the best interest of their pension rather than the best interest of their firm. One executive wanted to be remembered in history books as the pioneer of high-speed rapid transit. He came close to bankrupting his company in the process of achieving his personal ambitions at the expense of the projects he was sponsoring and also at the expense of the corporation.

Self-serving decision makers seem to focus on what is in their own best interest in the short term and often disregard what might be in the best interest of the project. In a project environment, this can become quite a complex process if the team members, stakeholders, the client, and the project sponsor all want the decision made in their own best interest. Suboptimal solutions are reached with several parties being quite unhappy with the final result. Unfortunately, self-serving decisions are almost always made for what is in the best interest of the largest financial contributor to the project for fear that, if the financial contributor removes support from the project, the project may be canceled.

10.11 DECISION MAKING: TOOLS AND METHODS

We all make decisions every day. We even use decision-making tools and do not realize it. Some of the decision-making tools and techniques people use in everyday life are:

- Determining the pros and cons for a given situation
- Choosing the alternative with the highest probability of occurrence
- Choosing the alternative that provides the greatest financial reward
- Choosing the alternative that offers the least amount of damage should something go wrong
- Accepting the first option that seems like it might achieve the desired result
- Following the advice of a subject matter expert
- Flipping a coin, cutting a deck of playing cards, and other random or coincidence methods
- Prayer, tarot cards, astrology, augurs, revelation, or other forms of divination

While these tools seem simplistic, there are more sophisticated tools that people use.

SWOT Analysis

There are numerous complex tools that can be used for more complex decision making tools. SWOT analysis looks at the strengths, weaknesses, opportunities, and threats in a given situation. SWOT analysis was originally created as a strategic planning tool but has now been adapted to complex problem solving on a project or in a business venture. It involves specifying the objective of the problem or project and identifying the internal and external factors that are favorable and unfavorable to achieve that objective.

A SWOT analysis must first start with defining a desired end state or objective for the problem at hand.

Strengths: Characteristics of the team that give it the ability to solve the problem. This could include technical knowhow and expertise.

Weaknesses: Characteristics that may prevent the team from solving the problem. This could be the team's lack of technical ability.

Opportunities: External opportunities if the problem is resolved.

Threats: External risks or elements in the environment or with stakeholders that could cause trouble for the project or business.

Strengths and weaknesses are *internal* strengths and weaknesses and look at the capability of the internal resources to solve the problem. Opportunities and threats are *external* results that may occur if the problem is or is not resolved. Strengths and weaknesses indicate what you "can do," and this must take place before you look at the opportunities and threats, which indicate what you "should do." Having a great alternative that will appease the stakeholders is nice as long as you have the qualified resources to accomplish it.

Pareto Analysis

Pareto analysis is a statistical technique in decision making that is used for selection of a limited number of tasks that produce significant overall effect such as the solution to a problem. It uses the Pareto principle—the idea that by doing 20% of work, 80% of the advantage of doing the entire job can be generated. In terms of quality improvement, a large majority of problems (80%) are produced by a few key causes (20%). In problem solving, 80% of the desired solution can be obtained by performing 20% of the work.

Pareto analysis is a formal technique useful where many possible courses of action are competing for the solution to a problem. In essence, the problem solver estimates the benefit delivered by each action, then selects a number of the most effective actions that deliver a total benefit reasonably close to the maximal possible one.

Pareto analysis is a creative way of looking at causes of problems because it helps stimulate thinking and organize thoughts. However, it can be limited by its exclusion of possibly important problems which may be small initially but grow with time. It should be combined with other analytical tools, such as failure mode and effects analysis and fault tree analysis.

This technique helps to identify the top 20% of causes of a problem that needs to be addressed to resolve the 80% of the problems. Once the top 20% of the causes are identified, then tools like the Ishikawa diagram or fishbone analysis can be used to identify the root causes of the problems.

Multiple-Criteria Decision Analysis

Multiple-criteria decision analysis (MCDA) or multiple-criteria decision making (MCDM) is a discipline aimed at supporting decision makers faced with making numerous and sometimes conflicting evaluations. MCDA aims at highlighting these conflicts during problem solving and deriving a way to come to a compromise. It is a combination of intuition and a systematic approach.

Unlike methods that assume the availability of measurements, measurements in MCDA are derived or interpreted subjectively as indicators of the strength of various preferences. Preferences differ from decision maker to decision maker, so the outcome depends on who is making the decision and what their goals and preferences are. Since MCDA involves a certain element of subjectivity, the morals and ethics of the persons implementing MCDA play a significant part in the accuracy and fairness of MCDA's conclusions. The ethical point is very important when one is making a decision that seriously impacts other people, as opposed to a personal decision.

There are many MCDA/MCDM methods in use today. However, often different methods may yield different results for exactly the same problem. In other words, when exactly the same problem data are used with different MCDA/MCDM methods, such methods may recommend different solutions even for very simple problems (i.e., ones with very few alternatives and criteria). The choice of which model is most appropriate depends on the problem at hand and may be to some extent dependent on which model the decision maker is most comfortable with. A question with all the above methods, and also methods not included in this list or even future methods, is how to assess their effectiveness.

Paired-Comparison Analysis

In paired-comparison analysis, also known as paired-choice analysis, decision alternatives are compared two at a time to see the relative importance. The alternatives are compared and the results are tallied to find an overall winner. The process begins by first identifying a range of plausible options. Each option is compared against each of the other options, determining the preferred option in each case. The results are tallied and the option with the highest score is the preferred option. This technique may be conducted individually or in groups. It may include criteria to guide the comparisons or be based on intuition following an open discussion of the group. A paired-choice matrix or paired-comparison matrix can be constructed to help with this type of analysis.

Decision Trees

A decision tree is a decision support tool that uses a treelike graph or model of decisions and their possible consequences, including chance event outcomes, resource costs, and utility. It is one way to display an algorithm for decision making. Decision trees are excellent tools for helping you to choose between several courses of action. They provide a highly effective structure within which you can lay out options and investigate the possible outcomes of choosing those options. They also help you to form a balanced picture of the risks and rewards associated with each possible course of action.

Decision trees are commonly used in operations research, specifically in decision analysis, to help identify the optimum approach most likely to reach a goal. Among decision support tools, decision trees (and influence diagrams) have several advantages:

- Are simple to understand and interpret. People are able to understand decision tree models after a brief explanation.
- Lay out the problem such that all options can be analyzed.
- Allow us to see the results of making a decision.
- Have value even with little hard data. Important insights can be generated based on experts describing a situation (its alternatives, probabilities, and costs) and their preferences for outcomes.
- Use a white box model. If a given result is provided by a model, the explanation for the result is easily replicated by simple math.
- Can be combined with other decision techniques such as probability trees.

Influence Diagrams

An influence diagram (ID) (also called a relevance diagram, decision diagram, or decision network) is a compact graphical and mathematical representation of a decision situation. It is a simple visual representation of a decision problem. Influence diagrams offer an intuitive way to identify and display the essential elements, including decisions, uncertainties, and objectives, and how they influence each other. It is a generalization of a Bayesian network, in which not only probabilistic inference problems but also decision-making problems (following maximum expected utility criterion) can be modeled and solved. Influence diagrams are very useful in showing the structure of the domain, that is, the structure of the decision problem. Influence diagrams contain four types of nodes (decision, chance, deterministic, and value) and two types of arcs (influences and informational arcs).

ID was first developed in the mid-1970s within the decision analysis community with an intuitive semantic that is easy to understand. It is now adopted widely and becoming an alternative to the decision tree, which typically suffers from exponential growth in number of branches with each variable modeled. ID is directly applicable in team decision analysis, since it allows incomplete sharing of information among team members to be modeled and solved explicitly. Extension of ID also finds its use in game theory as an alternative representation of the game tree.

As a graphical aid to decision making under uncertainty, it depicts what is known or unknown at the time of making a choice and the degree of dependence or independence (influence) of each variable on other variables and choices. It represents the cause-and-effect (causal) relationships of a phenomenon or situation in a nonambiguous manner and helps in a shared understanding of the key issues.

Affinity Diagrams

An affinity diagram is a technique for organizing verbal information into a visual pattern. An affinity diagram starts with specific ideas and helps you work toward broad categories. This is the opposite of a cause-and-effect diagram, which starts with the broad causes and works toward specifics. You can use either technique to explore all aspects of an issue. Affinity diagrams can help you:

- Organize and give structure to a list of factors that contribute to a problem.
- Identify key areas where improvement is most needed.

This technique is useful when there are large amounts of data. The affinity diagram is a business tool used to organize ideas and data. The tool is commonly used within project management and allows large numbers of ideas stemming from brainstorming to be sorted into groups for review and analysis.

The benefits include:

- Adding structure to a large or complicated issue
- Breaking down a complicated problem into broad categories
- Gaining agreement on the solution to a problem

Game Theory

Game theory models or games, as applied to project management problem solving and decision making, allow us to address a problem in which an individual's success in making choices depends on the choices of others. Simply stated, this technique considers responses of outside participants. It can be used to address how the client and stakeholders might react to certain alternatives selected.

It is used not only in project management but also in the social sciences (most notably in economics, management, operations research, political science, and social psychology) as well as in other formal sciences (logic, computer science, and statistics) and biology (particularly evolutionary biology and ecology). While initially developed to analyze competitions in which one individual does better at another's expense (zero sum games), it has been expanded to treat a wide class of interactions classified according to several criteria. This makes it applicable to project management, especially on complex projects where multiple stakeholders exist, each with competing needs.

Cost–Benefit Analysis

Cost–benefit analyses are most useful for problems involving financial decisions. The alternatives to a problem are usually those where the value of receiving the benefits outweighs the costs of obtaining them. Factors considered in cost–benefit analyses include:

- Return on investment
- Net present value
- Internal rate of return
- Cash flow
- Payback period
- Market share

Other parameters to consider that are more difficult to quantify include:

- Stockholder and stakeholder satisfaction
- Customer satisfaction
- Employee retention
- Brand loyalty
- Time-to-market
- Business relationships
- Safety
- Reliability
- Reputation
- Goodwill
- Image

Nominal Work Groups

- Work groups or nominal work groups, as applied to project management, can be an interdisciplinary collaboration of researchers or subject matter experts that have convened to identify and/or solve a problem. The group may be external consultants or contractors. The lifespan of the working group can be one day or several weeks. Such groups have the tendency to develop a *quasi-permanent existence* once the assigned task is accomplished; hence the need to disband (or phase out) the work group once it has provided solutions to the issues for which it was *initially* convened.
- The work group may assemble experts (and future experts) on a topic together for intensive work. It is not an avenue for briefing novices about the subject matter. Occasionally, a group might admit a person with little experience and a lot of enthusiasm. However, such participants should be present as observers and in the minority.
- It is imperative for the participants to appreciate and understand that the working group is intended to be a forum for cooperation and participation. Participants represent the interests and views of stakeholders from disparate sectors of the community which happen to have a vested interest in the solution to a problem. Therefore, maintaining and strengthening communication lines with all parties involved are essential (this responsibility cuts both ways—stakeholders are expected to share what information, knowledge, and expertise they have on the issue.)
- Each member of the work group may be asked to present their solutions to the rest of the group for analysis and be willing to accept constructive criticism. Work groups often have the advantage of arriving at a reasonably rapid decision but also suffer from the drawback that possibly not all of the alternatives were considered.

Delphi Techniques

The Delphi technique is a structured communication approach originally developed as a systematic, interactive forecasting and problem-solving method which relies on a panel of experts. The experts may not know who else is a member of the panel and all responses are provided anonymously.

In the standard version, the experts answer questionnaires in two or more rounds. After each round, a facilitator provides an anonymous summary of the experts' forecasts (or solutions to a problem) from the previous round as well as the reasons they provided for their judgments. Thus, experts are encouraged to revise their earlier answers in light of the replies of other members of their panel. It is believed that during this process the range of the answers will decrease and the group will converge toward the "correct" answer. If convergence does not take place, the panel may be asked to select the five best alternatives for the next round. Then, in the next round, select the three best alternatives. Then in the following round, select the two best alternatives. Finally, the process is stopped after a predefined stop criterion (e.g., number of rounds, achievement of consensus, and stability of results) and the mean or median scores of the final rounds determine the results.

Delphi is based on the principle that forecasts (or decisions) from a structured group of individuals are more accurate than those from unstructured groups. This has been indicated with the term "collective intelligence." The advantage of this technique is that the participants will provide their answers without being biased by others or openly criticized. People are free to state their opinion. The downside factors are that the process takes time (which may not be a luxury that most projects have) and that the best approach may be the combining of two or more alternatives rather than forcing people to select just one alternative.

Other Decision-Making Tools

There are other decision-making tools, some of which may take longer to perform. These include:

Linear Programming Applications: This includes the application of management science and operations research models for decision making.

Trial-and-Error Solutions: Useful for small problems when the cause-and-effect relationships are reasonable well known.

Heuristic Solutions: Similar to trial-and-error solutions but experimentation is done to reduce the list of alternatives.

Scientific Methods: This is used for problem solving involving scientific issues where additional experimentation may be done to confirm the problem and/or hypothesis

Problem-solving sessions normally involve only one decision-making tool. Most of the more complex tools are time consuming and costly to use, and combining several of these together may be prohibitive.

10.12 EVALUATING DECISION AND TAKING CORRECTIVE ACTION

Anybody can make a decision, but the hard part is making the right decision. Decision makers often lack the skills in how to evaluate the results or impact of a decision. What the project manager believes was the correct decision may be viewed differently by the client and the stakeholders.

Part of decision making requires the project manager to predict how those impacted by the decision will react. Soliciting feedback prior to the implementation of the solution seems nice to do, but the real impact of the decision may not be known until after full implementation of the solution. As an example, as part of developing a new product, marketing informs the project manager that the competition has just come out with a similar product and marketing believes that we must add some additional features into the product you are developing. The project team adds a significant number of "bells and whistles" to the point where the product's selling price is higher than that of the competition and the payback period is now elongated. When the product was eventually launched, the consumer did not believe that the added features were worth the additional cost.

It is not always possible to evaluate or predict the impact of a decision when making a choice among alternatives. But soliciting feedback prior to full implementation is helpful.

A useful tool for assisting in the selection of alternatives is a consequence table such as shown in Table 10-1. For each alternative, the consequences are measured against a variety of factors such as each of the competing constraints. For example, an alternative could have a favorable consequence on quality but an unfavorable consequence on time and cost. Most consequence tables have the impacts identified quantitatively rather than qualitatively. Risk is also a factor that is considered, but the impact on risk is usually defined qualitatively rather than quantitatively.

If there are three alternatives and five constraints, then there may be 15 rows in the consequence table. Once all 15 consequences are identified, they are ranked. They may be ranked according to either favorable or unfavorable consequences. If none of the consequences are acceptable, then it may be necessary to perform trade-offs on the alternatives. This could become an iterative process until an agreed-upon alternative is found.

The people preparing the consequence table are the people that make up the project team rather than possible outsiders that were brought in as subject matter experts for a particular problem. Project team members know the estimating techniques as well as the tools that are part of the organization process assets that can be used for determining impacts.

TABLE 10-1 Consequence Table

Alternative	Time	Cost	Quality	Safety	Overall Impact
#1	A	C	B	B	B
#2	A	C	A	C	B
#3	A	C	C	C	C
#4	B	A	C	A	B
#5	A	B	A	A	A

A= High impact
B= Moderate impact
C= Low impact

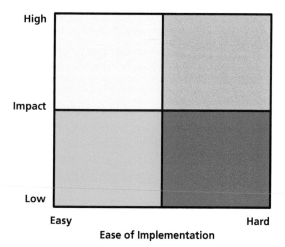

FIGURE 10-1   Impact analysis goal

It is nice to have several possible alternatives for the solution to a problem. Unfortunately, the alternative that is finally selected must be implemented, and that can create problems also.

One of the ways to analyze the impact is to create an impact–implementation matrix as shown in Figure 10-1. Each alternative considered could have a high or low impact on the project. Likewise, the implementation of each alternative could be easy or hard.

Each alternative is identified in its appropriate quadrant. The most obvious choice would be the alternatives that have a low impact and are easy to implement. But in reality, we often do not find very many alternatives in this quadrant.

Time to Implement Solution

We have all sat in problem-solving sessions and listened to team members come up with good (often brilliant) solutions to a problem. Everyone becomes enamored with the brilliance of the solution but nobody seems concerned as to how long it will take to implement the solution. Significantly more time is consumed in implementing a solution than in decision making.

Questions that should be considered include:

- Must we change our plans/baselines and, if so, how long will it take?
- How long will it take to get the necessary additional funding approved?
- Will the resources with the required skill levels be available when needed?
- Is overtime an option?
- How long will it take to procure those materials we need?
- Are additional reviews and meetings necessary before implementation can begin?
- Are additional reviews and meetings necessary as we implement the solution?
- Simply stated, decision making is easy; implementation is often difficult and time consuming.

DISCUSSION QUESTIONS

The discussion questions are for classroom use to stimulate group thinking about PM 2.0. There are no right or wrong answers to most of the questions.

1. What are the differences between problem solving and decision making?

2. Can some project managers be better at problem solving than decision making, and vice versa?

3. How can an executive find out during the selection of the project manager whether heor she has skills in problem solving and/or decision making?

4. How do creativity and innovation affect problem solving and decision making?

5. Who has the responsibility for approving the project manager's decision?

6. Can the alternatives for the solution to a problem come from the governance committee?

7. Can the alternatives for the solution to a problem come from the stakeholders?

8. How should a situation be handled where the stakeholders demand a specific solution that the project manager does not believe he or she can implement?

9. What would you consider as the best tools for problem solving?

10. What would you consider as the best tools for decision making?

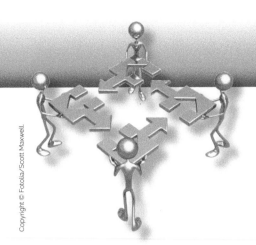

CHAPTER 11

NEED FOR PROJECT MANAGEMENT

11.0 BACKGROUND TO PROJECT MANAGEMENT MATURITY MODELS

As discussed in Chapter 1, a characteristic of PM 2.0 is a company's desire for organizational project management maturity. Maturity models were created to prepare companies for the future rather than the present. To understand this, you must first recognize what makes project management work well. Having an enterprise project management methodology does not necessarily lead to maturity. Having policies and procedures embedded throughout the methodology is also no guarantee that maturity will be forthcoming. Even following the *PMBOK® Guide* exactly cannot guarantee maturity.

Project management methodologies based upon rather rigid policies and procedures were created because management wanted standardization in the way that projects were planned, scheduled, and controlled. This was a necessity because executives had concerns about the ability of their project managers to make the correct decisions. The problem with standardization is that it often pulls people out of their comfort zones and they must work differently when assigned to projects. People who are asked to work outside of their comfort zone often dislike working on project teams and may look forward to the end of the project such that they can return to their previous assignment.

Project management excellence seems to come from four critical components:

- Effective communications
- Effective cooperation
- Effective teamwork
- Trust

With this in mind, most maturity models appear to be significantly more behavioral than quantitative. This is an important characteristic of PM 2.0. People manage projects; methodologies function as supporting tools. You can have the greatest methodology in the world and still not reach a level of maturity because the correct human behavior is not in place. Maturity in project management occurs when people work together correctly. The assessment instruments in maturity models should have some focus on people interacting with other people rather than just tools.

11.1 SOME BENEFITS OF USING A MATURITY MODEL

Over the years, executives have seen the benefits of using project management correctly. As executives demonstrated more trust in the project manager's capabilities, rigid methodologies were replaced with forms, guidelines, templates, and checklists. Today, at the beginning of a project, the project manager will walk through the "cafeteria" and select from the shelves only those forms, guidelines, templates, and checklists that are appropriate for that project and that client. If the project manager believes that this project is a very low risk, then the project manager may not want to follow or even use the risk management section of the *PMBOK® Guide*. Project managers are now being given more freedom on how to apply project management practices to satisfy the customer's needs. This leads to increased customer satisfaction and repeat business. This is one of the benefits expected with PM 2.0.

But even with this new freedom, the project managers must still recognize the importance of the behavioral maturity model assessments that focus on effective communication, cooperation, teamwork, and trust. Behavioral assessments indicate whether people believe that they are working within their comfort zone. If continuous improvements are made correctly and people are happy with their comfort zone, project management maturity can be achieved quickly. The focus on the maturity model in this chapter is that people manage projects; people manage tools; tools by themselves manage neither people nor projects. As a former Air Force lieutenant general stated, "You must never allow the tool to control the hand that's holding it." Maturity models should certainly include an assessment of whether the organization has the right tools in place. But in my opinion, the emphasis should be more on a behavioral assessment.

11.2 DETERMINING AMOUNT OF MATURITY NEEDED

The amount of maturity a company needs is customer driven rather than internally driven. Whenever a contractor allows its customer to become more mature than it is, very unfavorable results can occur. Among them, (1) the customer tells the contractor how the work should be done, (2) the customer may perform the work by themselves, and (3) the customer seeks out a more mature contractor during competitive bidding. Therefore, companies that rely heavily upon external customers for their revenue stream, such as project-driven companies, must NEVER allow their customers to achieve a greater degree of maturity than theirs. For these companies, project management maturity is a necessity for survival and the use of a maturity model is mandatory.

Companies today should be willing to perform a frequent self-assessment to make sure that the firm is continuously improving and reaching some level of maturity. During competitive bidding activities, customers are now asking the contractors to show how mature their organization is with regard to project management. Maturity assessments could be the difference between winning and losing a potential contract.

11.3 GETTING STARTED

It is always better to learn from the mistakes of others rather than from your own mistakes. There are several lessons that can be learned. First, strategic planning for project management maturity is essential, even a necessity. Without guidance from some sort of project management maturity (strategic planning) model, achieving maturity could take decades as you learn from mistakes. All project management maturity models are a form of strategic planning. Second, there must be a corporate commitment (especially at the executive levels) for maturity to occur, and the executives must see the "value" in achieving a defined level of maturity in a reasonable time frame. There are assessment questions on this in most maturity models. Third, there must exist a dedicated organization that drives the maturity process, and this normally becomes the responsibility of the PMO. Companies where PMOs take the lead in the assessment and continuous improvement processes generally reach levels of maturity quicker than those that do not have any involvement by the PMO.

11.4 THINGS CAN GO WRONG

Other than the behavioral issues I discussed before, there are several factors that prevent companies from achieving maturity. These include: (1) executives not seeing the value in project management or in project management maturity; (2) executives not recognizing that project management maturity is now a competitive weapon; (3) executives not realizing the importance of project management maturity to the customers and competitors; (4) executives not willing to establish a PMO to guide the maturity process; and (5) executives not willing to commit sufficient resources to achieving maturity. Obviously, there is a common theme in these five factors: EXECUTIVES. Hence, executive education has been a priority in recent years. Executives must see the return on investment as a result of using assessment instruments. There are assessment questions on executive expectations and involvement in maturity models. And, once again, this emphasizes the importance of behavioral assessment.

11.5 CHOOSING RIGHT MATURITY MODEL

There are several project management maturity models in the marketplace. And while they all have a different approach, they all have the same ultimate objective: maturity! The decision of which model is best for a given company might be based on the time frame allotted, number of resources available, pressure by the customers, maturity level of the competitors, and whether the company is project driven or non–project driven.

Today, there are numerous papers published as well as Master's and Ph.D. theses that benchmark the various models. Even though I am somewhat partial to the model used in this chapter, there are other models for maturity assessments. What should be important is not necessarily what model you select but the fact that you are doing an assessment. In my opinion, all of the models in the marketplace provide some type of value if used properly.

There are two primary components that must be considered: simplicity and assessment capability. Published articles on maturity model benchmarking may have dozens of components, many of which are industry specific, but these two components are essential as starters. The prospect of using a complex maturity model may very likely scare away senior management because they may not be able to determine the time frame or resources needed to achieve maturity. With maturity models, complexity breeds avoidance. With regard to capability, assessment instruments are needed to identify areas of improvement and show that progress is being made and that continuous improvements in project management are adding value to the business.

11.6 ESTIMATING TIME TO REACH MATURITY

Experience has shown that the single most important force for achieving higher levels of maturity (other than continuous executive support) is the early-on establishment of a PMO. The PMO becomes the major driver for the maturity process. Without a PMO, it may take three to five years to reach certain initial levels of maturity. With early establishment of a PMO, however, and the right people assigned to the PMO, it may take only two years. The problem with deciding upon a time frame for maturity is heavily based upon someone's definition of maturity, the speed with which tools are either purchased or developed, and the commitment to the right levels of project management education.

Any organization can develop all the tools necessary to achieve maturity. But if the organization does not understand the benefits and value of project management or the use of the tools, what has it really accomplished? Maturity IS NOT the development of tools or processes. Maturity is the effective USE of these instruments and continuous improvement in use of these instruments using captured best practices. Whenever companies ask whether or not the investment of time and money to obtain maturity is worth it, the response is simple. You know the amount of money needed to achieve a certain level of maturity. But what is the cost or opportunity loss of not achieving it? It is possible for the opportunity loss to be at least an order of magnitude greater than the cost of achieving maturity.

11.7 STRATEGIC PLANNING FOR PROJECT MANAGEMENT MATURITY

All companies desire excellence in project management. Unfortunately, not all companies recognize that the time frame can be shortened by performing strategic planning for project management. The simple use of project management, even for an extended period of time, does *not* lead to excellence. Instead, it can result in repetitive mistakes and, worse, learning from your own mistakes rather than from the mistakes of others.

Strategic planning for project management is unlike other forms of strategic planning in that it is most often performed at the middle-management level, rather than by executive management. Executive management is still involved, mostly in a supporting role, and provides funding together with employee release time for the effort. Executive involvement will be necessary to make sure that whatever is recommended by middle management will not result in unwanted changes to the corporate culture.

Organizations tend to perform strategic planning for new products and services by laying out a well-thought-out plan and then executing the plan with the precision of a surgeon. Unfortunately, strategic planning for project management, if performed at all, is done on a trial-by-fire basis. However, there are models that can be used to assist corporations in performing strategic planning for project management and achieving maturity and excellence in a reasonable period of time.

11.8 PROJECT MANAGEMENT MATURITY MODEL

The foundation for achieving excellence in project management can best be described as the project management maturity model (PMMM), which is comprised of five levels, as shown in Figure 11-1. Each of the five levels represents a different degree of maturity in project management.

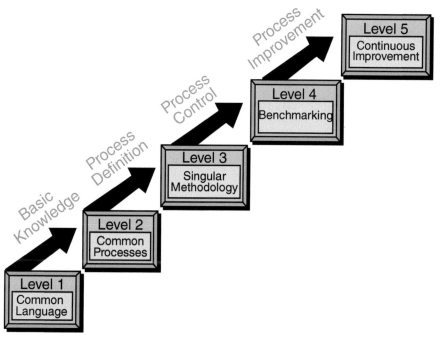

FIGURE 11-1 **Five levels of maturity**

Level 1—Common Language: In this level, the organization recognizes the importance of project management and the need for a good understanding of the basic knowledge on project management, along with the accompanying language/terminology.

Level 2—Common Processes: In this level, the organization recognizes that common processes need to be defined and developed such that successes on one project can be repeated on other projects. Also included in this level is the recognition that project management principles can be applied to and support other methodologies employed by the company.

Level 3—Singular Methodology: In this level, the organization recognizes the synergistic effect of combining all corporate methodologies into a singular methodology, the center of which is project management. The synergistic effects also make process control easier with a single methodology than with multiple methodologies.

Level 4—Benchmarking: This level contains the recognition that process improvement is necessary to maintain a competitive advantage. Benchmarking must be performed on a continuous basis. The company must decide whom to benchmark and what to benchmark.

Level 5—Continuous Improvement: In this level, the organization evaluates the information obtained through benchmarking and must then decide whether or not this information will enhance the singular methodology.

When we talk about levels of maturity (and even life-cycle phases), there exists a common misbelief that all work must be accomplished sequentially (i.e., in series). This is not necessarily true. Certain levels can and do overlap. The magnitude of the overlap is based upon the amount of risk the organization is willing to tolerate. For example, a company can begin the development of project management checklists to support the methodology while it is still providing project management training for the workforce. A company can create a center for excellence in project management before benchmarking is undertaken.

Although overlapping does occur, the order in which the phases are completed cannot change. For example, even though level 1 and level 2 can overlap, level 1 *must* still be completed before level 2 can be completed. Overlapping of several of the levels can take place, as shown in Figure 11–2.

Overlap of Level 1 and Level 2: This overlap will occur because the organization can begin the development of project management processes either while refinements are being made to the common language or during training.

Overlap of Level 3 and Level 4: This overlap occurs because, while the organization is developing a singular methodology, plans are being made as to the process for improving the methodology.

Overlap of Level 4 and Level 5: As the organization becomes more and more committed to benchmarking and continuous improvement, the speed by which the organization wants changes to be made can cause these two levels to have significant overlap.

The feedback from level 5 back to level 4 and level 3, as shown in Figure 11-3, implies that these three levels form a continuous improvement cycle, and it may even

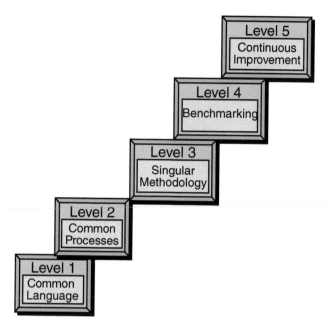

FIGURE 11-2 Overlapping levels

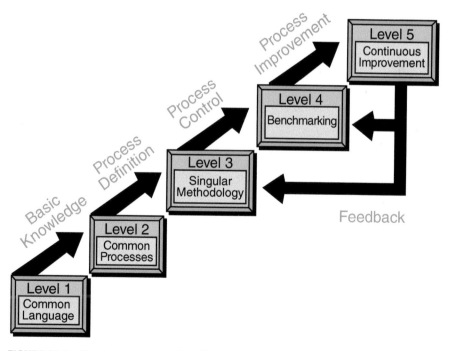

FIGURE 11-3 Five levels of maturity with a feedback loop

TABLE 11-1 Degrees of Difficulty per Level		
Level	**Description**	**Degree of Difficulty**
1	Common language	Medium
2	Common processes	Medium
3	Singular methodology	High
4	Benchmarking	Low
5	Continuous improvement	Low

be possible for all three of these levels to overlap. Level 2 and level 3 generally do not overlap. It may be possible to begin some of the level 3 work before level 2 is completed, but this is highly unlikely. Once a company is committed to a singular methodology, work on other methodologies generally terminates. Also, companies can create a Center for Excellence in project management early in the life-cycle process but will not receive the full benefits until later on.

Risks can be assigned to each level of the PMMM. For simplicity's sake, the risks can be labeled as low, medium, and high. The level of risk is most frequently associated with the impact on the corporate culture. The following definitions can be assigned to these three risks:

Low Risk: Virtually no impact upon the corporate culture or the corporate culture is dynamic and readily accepts change.

Medium Risk: The organization recognizes that change is necessary but may be unaware of the impact of the change. Multiple-boss reporting would be an example of a medium risk.

High Risk: High risks occur when the organization recognizes that the changes resulting from the implementation of project management will cause a change in the corporate culture. Examples include the creation of project management methodologies, policies, and procedures as well as decentralization of authority and decision making.

Level 3 has the highest risk and degree of difficulty for the organization. This is shown in Table 11-1. Once an organization is committed to level 3, the time and effort needed to achieve the higher levels of maturity have a low degree of difficulty.

Achieving level 3, however, may require a major shift in the corporate culture.

11.9 PM 2.0 INPUT INTO PMMM

Maturity models need to be updated based upon changes in project management. The growth of PM 2.0 has mandated four changes to be made to PMMM:

- An understanding of how metrics and metric management programs impact the maturity process (see Figure 11-4)
- An understanding of how committee governance impacts the maturity process (see Figure 11-5)
- An understanding of how the addition of value as a success criterion and possibly a constraint impacts the maturity process (see Figure 11-6)
- An understanding of how changes in leadership, such as needed on transformational projects, impacts the maturity process (see Figure 11-7)

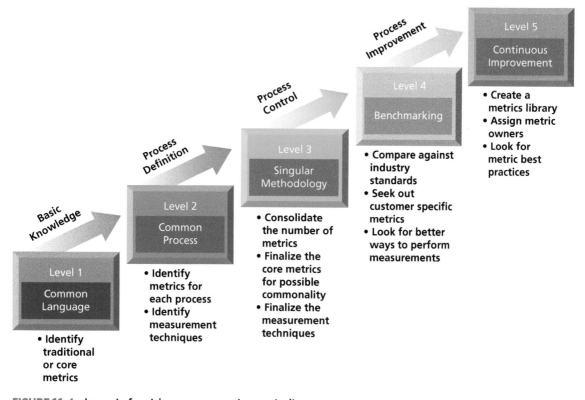

FIGURE 11-4 Impact of metrics management on maturity

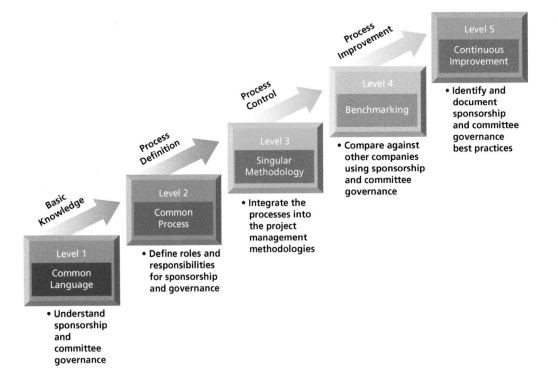

FIGURE 11-5 Impact of governance on maturity

These types of maturity models will become more common in the future, with generic models being customized for individual companies. These models will assist management in performing strategic planning for excellence in project management.

DISCUSSION QUESTIONS

The discussion questions are for classroom use to stimulate group thinking about PM 2.0. There are no right or wrong answers to most of the questions.

1. What is the definition of project management maturity?
2. Can project management maturity ever be reached?
3. Why are there so many project management maturity models in the marketplace?
4. What factors determine which maturity model is best for a given company?
5. Would you expect continuous improvement efforts to be part of all project management maturity models?

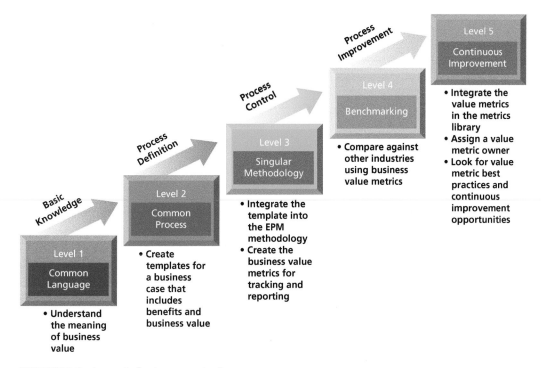

FIGURE 11-6   Impact of value on maturity

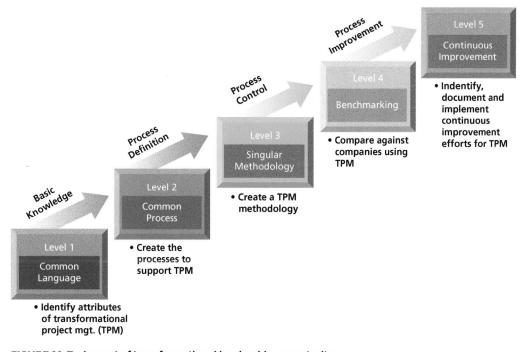

FIGURE 11-7   Impact of transformational leadership on maturity

6. What can cause a project management maturity model to fail or give improper results?

7. Who should assume the leadership role for assessing project management maturity?

8. How long should it take to reach some degree of maturity, and what assumptions have you made?

9. What should be the role of senior management during project management maturity efforts?

10. During competitive bidding, should a customer have the right to expect a certain level of maturity?

CHAPTER 12

USING THE PMO TO SPEARHEAD PM 2.0[1]

12.0 INTRODUCTION

Someone in the organization must take the responsibility for serving as the champion for the implementation of PM 2.0. This will fall on the shoulders of the PMO. There are several types of PMOs that can take the lead, although the implementation of PM 2.0 practices might be easier in certain PMOs. But regardless of which PMO takes the lead, there will be roadblocks.

Today, companies are managing their business by projects. The result has been a vast amount of project management information surfacing from all areas of the company. This information focuses on best practices in the project management, the usefulness of an enterprise project management methodology that has some degree of flexibility, the benefits of project management, how to best implement improvements in project management, and how project management is benefiting the profitability of the company. As companies begin to recognize the favorable effect that project management has on performance, all of this project management knowledge is treated as intellectual property. Emphasis is now placed upon achieving professionalism in project management using the project office (PO) or the PMO that functions as the guardian for the project management intellectual property. The concept of a PO or PMO could very well be the most important project management activity in this decade.

12.1 TRADITIONAL PROJECT OFFICE

In the early years of project management, large projects were managed by a PO which was composed of the project manager and assistant project managers. The project

---

[1] Some of this chapter has been adapted from H. Kerzner, *Project Management: A Systems Approach to Planning, Scheduling, and Controlling*, 11th ed., Wiley, Hoboken, NJ, 2013, Chapter 23.

office as it was called then was committed to just one and only one project and therefore focused on one and only one customer.

All POs had to follow the same project management methodology, but each PO could have unique policies and procedures related to servicing the one client. All changes to the standard policies and procedures had to be approved by the project sponsor. On some projects, the personnel assigned to the PO could not communicate directly with the client without first going through the project sponsor.

Some best practices were captured, but these related specifically to interfacing with one client. Even if the best practices were opportunities that could benefit other projects, the information was rarely exchanged because the POs did not talk to one another. The POs often competed with one another for the same resources and decisions were made for the best interest of the project rather than the best interest of the company as a whole.

12.2 TRADITIONAL PMO

Over time, the benefits of using project management, which were once seen as applicable only to the aerospace, defense, and heavy construction industries, were now recognized as being applicable for other industries. By the late 1990s, as more of the benefits of project management became apparent, management understood that there might be a significant, favorable impact on the corporate bottom line of the entire company rather than just one project. This led management to two important conclusions:

- Project management must be recognized as a career path position.
- There must be a centralized group within the company to promote continuous improvement activities in project management.

The second bullet led to the creation of the PMO concept. While the traditional PO was dedicated to a single project, the PMO was dedicated to the entire company and had to support all projects. Consideration was being given for all critical activities related to project management to be placed under the supervision of the PMO. This included such topics as:

- Standardization in estimating
- Standardization in planning
- Standardization in scheduling
- Standardization in control
- Standardization in reporting
- Clarification of project management roles and responsibilities
- Preparation of job descriptions for project managers
- Preparation of archive data on lessons learned
- Continuous project management benchmarking
- Developing project management templates
- Developing a project management methodology
- Recommending and implementing changes and improvements to the existing project management methodology
- Identifying project management standards

- Identifying best practices in project management
- Performing strategic planning for project management
- Establishing a project management problem-solving hotline
- Coordinating and/or conducting project management training programs
- Transferring knowledge through coaching and mentorship
- Developing a corporate resource capacity/utilization plan
- Assessing risks in projects
- Planning for disaster recovery in projects
- Performing or participating in the portfolio management of projects
- Acting as the guardian for project management intellectual property

With these changes taking place, some organizations began changing the name of the PMO to Center of Excellence (COE) in project management. The COE was mainly responsible for providing information to internal stakeholders rather than actually executing projects or making midcourse corrections to a plan. The PMO was seen as overhead rather than as a direct labor charge against projects. The PMO's mission was planning for the future of project management rather than the profitability of individual projects.

12.3 IMPLEMENTATION RISKS

Each activity assigned to the PMO brought with it both advantages and disadvantages. The majority of the disadvantages were attributed to the increased levels of resistance to the new responsibilities given to the PMO. Much of the resistance was at the executive levels where many executives were afraid that the executive who had control of the PMO and all of this intellectual property might become more powerful than other executives. Another fear was that the PMO might establish project management policies and procedures and then force the executive's functional group to change the way that they conduct business.

For simplicity sake, the resistance levels can be classified as low risk, moderate risk, and high risk according to the following definitions:

- **Low Risk:** Easily accepted by the organization with very little shift in the balance of power and authority. Virtually no impact on the corporate culture.
- **Moderate Risk:** Some resistance by the corporate culture and possibly a shift in the balance of power and authority. Resistance levels can be overcome in the near term and with minimal effort.
- **High Risk:** Heavy pockets of resistance exist and a definite shift in some power and authority relationships. Strong executive leadership may be necessary to overcome the resistance.

Not every PMO has the same responsibilities. Likewise, the same responsibilities implemented in two PMOs can have differing degrees of the best interest of the organization. Evaluating potential implementation risks is critical. It may be easier to gain support for the establishment of a PMO by implementing low-risk activities first. The low-risk activities are operational activities to support project management efforts in the near term whereas the high-risk activities are more in line with strategic planning responsibilities and possibly the control of sensitive information. For example, low-risk activities include mentorship, developing standards, and project management training. High-risk activities include capacity planning, benchmarking, and dissemination of information.

Senior managers were now recognizing that project management and the PMO had become invaluable assets for senior management as well as for the working levels.

The benefits for the executive levels of management of using a PMO had become apparent. They included:

- Standardization of operations
- Company rather than silo decision making
- Better capacity planning (i.e., resource allocations)
- Quicker access to higher quality information
- Elimination or reduction of company silos
- More efficient and effective operations
- Less need for restructuring
- Fewer meetings that historically robbed executives of valuable time
- More realistic prioritization of work
- Development of future general managers

All of the above benefits were either directly or indirectly related to the project management intellectual property. To maintain the project management intellectual property, the PMO must maintain the vehicles for capturing the data and then disseminate the data to the various stakeholders. These vehicles include the company project management intranet, project websites, project databases, and project management information systems. Since much of this information is necessary for both project management and corporate strategic planning, there must exist strategic planning for the PMO.

PMOs became commonplace in the corporate hierarchy. Although the majority of activities assigned to the PO had not changed, there was now a new mission for the PO: supporting the entire corporation. The PO was now servicing the corporation, especially the strategic planning activities, rather than focusing on a specific customer or just project management. The PO was transformed into a corporate center for control of the project management intellectual property. This was a necessity as the magnitude of project management information grew almost exponentially throughout the organization.

12.4 SPECIALIZED PMO

Many executives felt threatened by the PMO concept. Not only was there a fear of how powerful a PMO could become, there was also the fear that the PMO might begin making decisions that were normally made by executives.

As the resistance to the PMO concept grew, some firms created functional PMOs. This type of PMO was utilized in one functional area or division of an organization, such as information systems. The major responsibility of this type of PMO was to manage a critical resource pool, that is, resource management. The PMO may or may not actually manage projects but may make decisions on how resources should be assigned to other projects. Another issue is that IT personnel have different life-cycle phases in their systems development methodologies than used in enterprise project management methodologies and argue that many of the practices developed by the traditional PMO should not be forced upon them.

Another type of specialized PMO is the customer group PMO. This type of PMO was created for better customer management and customer communications when managing several projects for the same customer. Common customers or projects were clustered together for better management and customer relations. Multiple customer group PMOs can exist at the same time and may end up functioning as a temporary organization. In effect, this acts like a company within a company. This type of PMO had a permanent project manager assigned and managing projects.

12.5 STRATEGIC PMO

This type of PMO services the entire company and focuses on corporate and strategic issues rather than functional issues but usually is not considered a specialized PMO. This is the type of PMO that would assume the responsibility for the overall implementation of PM 2.0 activities although other types of PMOs could implement pieces of PM 2.0. If this type of PMO does manage other types of projects, it is for cost reduction efforts.

The strategic PMO can also support portfolio management efforts or there may be a separate portfolio PMO established. Both the portfolio and strategic PMOs have the responsibility of aligning traditional project management practices with ongoing business practices. This is a necessity for PM 2.0 implementation. We stated in earlier chapters that project managers are now expected to make both project and business decisions. Also, most enterprise project management methodologies now contain business processes as well as project management processes.

Some companies have found success in maintaining a portfolio PMO dedicated exclusively to managing the portfolio of projects. This is also a characteristic of PM 2.0. This includes capacity-planning activities, project prioritization activities, and providing recommendations to senior management for project selection. The PMO also audits the ongoing portfolio of projects and may even perform health checks. The benefits of the portfolio PMO are relatively clear. The PMO enables businesses to[2]:

- Provide a structure for selecting the right projects and eliminating wrong ones
- Allocate resources to the right projects, thus reducing wasteful spending
- Align portfolio decisions to strategic business goals
- Base portfolio decisions on logic, reasoning, and objectivity
- Create ownership among staff by involvement at the right levels
- Establish avenues for individuals to identify opportunities and obtain support
- Help project teams understand the value of their contributions

[2] J. Pennypacker and S. Retna, Eds., *Project Portfolio Management: A View from the Management Trenches*, The Enterprise Portfolio Management Council, Wiley, Hoboken, NJ, 2009, p. xvi.

12.6 NETWORKING PMOS

Because of political infighting for control of the PMOs, many companies have established multiple PMOs all of which are networked together by a "coordinating" or master PMO. Other companies that are multinational have created regional PMOs that are groupings of project management associates (project managers, team members, etc.) who perform project management duties within specific regional or industry-specific areas. In this case, the master PMO responsibilities are:

- Promoting the enterprise project management methodology
- Promoting the use of standard project management tools
- Assuring standardization in project execution and delivery
- Maintaining a source of project management subject matter expertise
- Coordinating multinational project management knowledge

12.7 TRUST OF PROJECT GOVERNANCE

For most companies, project management began with the creation of a project management methodology that was based upon rather rigid policies and procedures. Every part of the methodology had to be used on all projects even if that particular step was not necessary. Senior management did not trust the project managers with making correct decisions and therefore used the methodology to provide standardization and maintain control over how the projects would be managed. Most project managers had limited freedom to depart from the methodology. Sponsorship was performed at middle management levels rather than senior management levels because senior management was fearful that their career could be damaged if they served as a sponsor and either the project failed or project management did not work.

As project management began to grow and project successes appeared, more trust was given to project managers. However, as the concept of the PMO developed, the real trust was given to the PMO. Project managers were trusted to make decisions in the best interest of the project, but the PMO had to validate that the decisions were also in the best interest of the company.

One of the prime characteristics of PM 2.0 is that project managers are given significantly more freedom in how to manage the project and to make the necessary decisions, within limits. However, regardless of the trust, there will still be governance on all projects, programs, and portfolio activities. Based upon the type of PMO, governance could appear at the seniormost levels of management. Also, companies began trusting the PMO to make better assessment of various project risks than the project managers.

As seen in Table 12-1, the level of project sponsorship and risk management activities appear to be different based upon the type of PMO.

Today, as our projects tend to become larger and more complex, sponsorship will be accomplished by a committee that will have the expertise in evaluating business risks. Project managers will still participate in the risk management assessments but may not have all of the necessary information related to business risks.

TABLE 12-1 PMO Governance and Risks			
Activity	**Traditional PMO**	**Strategic PMO**	**Portfolio PMO**
Project sponsorship	Middle management sponsorship but occasionally senior management. The PMO may assist in decision making and audits for methodology compliance but not sponsorship.	Business unit manager, but possibly higher up based upon the risk of the project. PMO provides little sponsorship but may perform audits and risk management assistance.	Almost all sponsorship done using committee governance with membership from each business unit. Risk management performed almost entirely by the PMO.
Sponsorship involvement	Decisions made in the best interest of a single project. Sponsorship done on an as-needed basis.	Best interest of the project and its alignment to the business objectives. Sponsorship on an as-needed basis.	Frequent if not continuous governance.
Risk management involvement	Mainly done by the project manager for technical risk and done by the sponsor for business risks.	Mainly business risks of the projects and the alignment to strategic objectives.	Almost exclusively done by the governance committee and the PMO.
Project crisis management	PMO involvement if requested.	Some business unit involvement as needed.	Continuous involvement by the PMO.

12.8 WAYS A PMO CAN FAIL[3]

Deciding to implement a PMO is easy. Being able to implement the decision, however, may be difficult because of all of the roadblocks that could exist. As companies recognize the need for project management, the need for the centralization of project management knowledge becomes apparent. Unfortunately, companies tend to focus on the good things that can happen without fully understanding the roadblocks that can impede successful implementation.

Simply stated, "Be careful what you wish for."

Unclear Mission Statement

There must be a valid reason why a company wants to establish a PMO. All too often, the PMO is established to either appease project managers or in the belief that the company must follow an example set by their competition. This is the result of a lack of vision on the benefits of a PMO.

Regardless of the reason, there must be a clearly understandable mission statement signed off by one or more senior managers. An unclear mission statement encourages

---

[3] Adapted from D. A. Dell, PMP®, and H. Kerzner, "Ways that a PMO Can Fail," *allPM*, April 14, 2014. Deborah Dell is the Director, IBM Project Management Center of Excellence.

failure. People may not understand the role of the PMO. Likewise, PMO members may not understand their own roles and responsibilities and end up working on activities that may have no direct relationship to corporate objectives. All of this can result in a lack of direction. For example, the PMO members may assign themselves work on projects that are self-serving rather than supporting the overall business. Or, they may commit vast resources for the support of a single project and disregard other projects that may have a greater need for their services or can provide a much greater value to the business.

Mission statements must be future oriented rather than focusing on the here-and-now. Typical goals for mission statements can include:

Continuous Improvements to Project Management Processes: Without continuous improvement, the organization can become complacent and miss opportunities.

Opportunity Identification: The PMO must help the organization achieve its strategic goals. Without opportunity identification, the organization worries about today rather than tomorrow.

Change Management Champions: Opportunity identification and continuous improvement most frequently require organizational change. Unless the PMO has the necessary skills to take the leadership role in organizational change, the change may never occur and the PMO fails to achieve its mission.

Failing to Focus on Impact to Business

The mission statement for the PMO should be to serve the overall business needs, not merely the needs of selected projects. Therefore, the PMO must be willing to make decisions that are in the best interest of both the business and the projects, all aligned with business strategy and business objectives. While it is true that some PMOs are more operational than strategic, the alignment of business decisions with business strategy must still exist.

The PMO is expected to add business value to the company. If the company cannot recognize the business value created by the PMO, then the PMO has failed and will most likely be disbanded.

Failing to Gain Implementation Support

Deciding to implement a PMO is a lot easier than gaining support during implementation. There are several activities that can be assigned to a PMO. A partial list of activities might be:

- Strategic planning for project management
- Benchmarking internally and externally
- Cost reduction and continuous improvements efforts
- Mentorship for new project managers
- Capturing lessons learned and best practices
- Maintaining a problem-solving hotline
- Creating templates
- Assisting Human Resources (HR) in developing a project management career path

- Supporting education and training for project management
- Assisting projects with disaster recovery plans
- Assessing risks
- Supporting customer relations management activities
- Supporting executives during portfolio selection and management
- Capacity planning efforts
- Maintaining a project management information system (PMIS)
- Guardian of project management intellectual property

The order in which these activities are implemented is important. For example, if the first activity assigned to the PMO is to support strategic planning activities, even if just related to project management, some executives may feel threatened by the existence of the PMO with the belief that the PMO will now be responsible for some of their efforts. The result may induce a lack of support for the PMO. It is therefore essential that PMO activities be assigned in the correct order such that support will be forthcoming. Also, trying to do too much too soon can lead to PMO failure.

Determining PMO Head Count

It is very easy to overstaff a PMO. Several years ago, a company determined the need to implement project management and created a project management methodology. The company was losing contracts to its competitors and the need to become good at project management in the shortest amount of time was essential.

The company created a PMO with a staff of more than 50 people. The majority of the people were transferred from business units to the PMO. The business unit managers eventually filled all of the vacated slots.

Three years later, the company had created a project management methodology and believed that they were now competitive in the marketplace. The PMO was treated as a cost center where the salaries of the employees in the PMO were part of the overhead structure of the company. Unfortunately, after economic conditions in the marketplace became poor, executives began looking for ways to reduce costs, beginning of course with the overhead structure of the company. The decision was made to reduce the head count in the PMO to less than 15 people. Unfortunately, there were no positions available in the company where these employees could be reassigned and they were eventually terminated. People began believing that an assignment to the PMO was not a career path opportunity and the support for the PMO diminished.

While we tend to focus on overstaffing a PMO, there is also the risk of understaffing the PMO. If the PMO is assigned more work than it can handle, then the effectiveness of the PMO will be the subject of discussion and the PMO may be eventually dissolved. This will most certainly occur if the PMO is unable to provide the value that is expected from it.

Failure of Success

Several years ago, an auto supplier in Detroit established a PMO with a mission statement that included the creation of a world-class methodology for automotive products. The PMO accomplished its task and the number of project successes grew. The PMO eventually became complacent and began focusing on getting people to use the methodology. Three years later, senior management began asking what continuous improvement efforts took place over the past three years. Much to their surprise, no improvements had taken place.

The organization had become complacent. The PMO lost its vision of the future and spent three years worrying heavily about the present. When significant successes occurred, it became a natural tendency to avoid "rocking the boat" and continue doing exactly what it had done in the past.

When a PMO becomes stagnant, executives wonder why the PMO still exists. Most PMOs are overhead rather than direct labor and, as such, are seen as opportunities for cost reduction. When the PMO becomes stagnant, people may believe it has accomplished its mission and should be shut down.

Failing to Use Intellectual Property Correctly

Project management intellectual property consists of proper understanding of the organizational process assets, the capturing and further use of best practices, an understanding of information stored in the knowledge repository systems for project management, and taking advantage of project management educational opportunities. This intellectual property is designed to improve one's project management skills by learning from what others have done.

There are two, and only two, ways by which an individual can become good at project management: you can learn from your own mistakes or from the mistakes of others from the intellectual property provided to you. Learning from your own mistakes can be a tedious and costly process. The organization may be willing to allow some projects to fail, either partially or totally, in order for learning to take place. This is somewhat unfortunate, but it was a way of life in the early years of project management, well before the existence of PMOs. Project management knowledge is intellectual property that firms must use effectively.

The creation of a project management methodology, accompanied by forms, guidelines, templates, and checklists, has the benefit of providing standardization and repeatability but does not necessarily provide project management education. Rather, it simply provides a roadmap. Educational opportunities must be provided to the workers.

There are several ways that project management learning can take place: college/university coursework, onsite specialized training, knowledge transfer, and certification programs. Effective PMOs are actively involved in onsite specialized training, knowledge transfer, and certification programs. The PMO must make sure that the course content and possibly the examples provided by the instructors are applicable to the company and/or industry. Otherwise, the workers may respond that "the information was nice to know but doesn't apply to us." Some companies have certification boards which must approve all certification programs to guarantee a good fit with the firm's educational goals.

Most companies are willing to provide their workforce with educational opportunities. In project management, it is the responsibility of the PMO to make sure that the correct project management education is provided so there are benefits for both the firm and the workers.

Failing to Collect Intellectual Property

Knowledge transfer comes from the collection of lessons learned and best practices. It can be discovered from both successes and failures. The PMO is the company's guardian of project management intellectual property and must make sure that the information is both gathered and correctly disseminated to the workforce. Even though there may be best practice owners assigned throughout the company, as well as subject matter experts to evaluate whether something is a best practice, someone from the PMO must accept the leadership role for the collection and disposition of the intellectual property.

When best practices are not managed correctly, people end up learning from their own mistakes rather than from the mistakes of others. Mistakes are often repeated, not only by individuals within the same division, but also by individuals in sister divisions. The failure of the PMO to collect intellectual property can significantly impede or even prevent continuous improvement efforts from occurring.

Forcing Organizational Change

The capturing of best practices and lessons learned often leads to organizational change. The PMO functions as the leader for that change. Not all changes are necessary and some changes that are necessary may be delayed until sometime in the future. Forcing unnecessary organizational change or dealing with a culture that tends to resist change can cause the PMO to fail.

Not Understanding Culture and Needs

People tend to dislike changes to their comfort zone. In such cases, it is not uncommon for people to bad-mouth the PMO and recommend it be dissolved. Before recommending any type of change, the PMO must understand the culture and the needs of the individuals. Most people seem to understand the need for change, but it is the how and when that people are concerned about.

In one company, the PMO implemented changes to the project management methodology that resulted in several decisions that were normally made by some of the blue collar workers to now be made by the leadership of the projects. The PMO was unaware that having the authority to make at least some decisions, no matter how small, could be seen as an opportunity for some blue collar workers to advance. Removing this opportunity from the blue collar workforce resulted in a lack of support for the PMO.

In another company, the PMO was asked to create a monetary award system for people working on project teams. The PMO put the system into action without considering the potential impact on the blue collar wage and salary administration program. Some blue collar workers assigned to project teams eventually earned more money than their colleagues that were at a higher pay grade because of project bonuses. Now, blue collar workers were fighting to work on project teams rather than performing their routine non–project work requirements.

Making the PMO a Profit Center

Several years ago, a company with a well-established PMO had a mission statement that focused heavily upon the future. All of the project managers were assigned to business units but reported on a "dotted line" basis to the PMO for the sharing of intellectual property and continuous improvement efforts. The head of each business unit had overall responsibility for profit and loss (P&L) for the business unit and this P&L responsibility was subsequently handed down to each of the project managers. Business units were treated as profit centers but the PMO was regarded as a cost center.

The company then hired a new president who decided that all of the project managers had to be centralized and with "solid line reporting" to the PMO. The project managers were still required to maintain P&L responsibility for each project. This act converted the PMO from a cost center to a profit center. The PMO now focused on the profitability of projects.

The results were devastating. Long-term continuous improvement efforts were avoided in favor of short-term efforts to increase profitability. Project management education was viewed as an unnecessary expense. The PMO had lost its identity. While rank has its privileges, and the president can restructure any way he or she desires, making the PMO a profit center resulted in the failure of the PMO.

Focusing on Profitability of a Single Project

While it is a fact that most PMOs do not have the responsibility for P&L, they are still expected to support all projects equally for the maximization of corporate profitability. The PMO can fail if it overemphasizes the profitability and support for one project at the expense of all of the other projects. This often occurs when a project "cries for help." PMO employees may also shirk some of their other duties to support just one project. Supporting a project in trouble is certainly the right thing to do, but restraints should be imposed on the amount and level of support provided.

Improper Staffing of PMO

There are not very many courses in the marketplace designed to train people in how to work in a PMO. People are expected to bring with them the necessary skills when assigned to a PMO. Some people believe that the three most important skills are process skills, communication skills, and project management skills. Too often people accept temporary assignments to a PMO or are permanently assigned to a PMO without fully understanding the roles and responsibilities of the PMO. Sometimes, the executives that support the existence of the PMO provide ill-defined requirements for roles and responsibilities, resulting in the wrong people being assigned to the PMO.

Project managers are quite adept at speaking in project management lingo. This is acceptable when managing a project and interfacing with project personnel, but when assigned to a PMO, you must be able to communicate with everyone and not all of the people you communicate with will have your understanding of project management lingo.

Unaware of Existence of PMO

It is very hard for people to support a PMO if they are unaware of its existence. This is particularly true of divisions that are remotely located away from the organization that houses the PMO. Even if the existence of it is known, there can still be a lack of support if the people do not understand why the PMO was created and what the responsibilities are for the PMO.

Some companies that are multinational firms may have multiple PMOs worldwide. In such a case, one PMO is considered as the "master" that has the responsibility of networking together all of the other PMOs. Care must be taken that each PMO understands its own responsibilities. For example, the master PMO may be responsible for continuous improvements to the methodology whereas the regional PMOs may be responsible for implementation of the changes.

Because the PMO is the guardian of project management intellectual property, some executives may feel threatened by PMOs that are directly under the control of other executives. Information is perceived as power and sharing information with PMOs that are not under your control could make other executives more powerful than you. Sometimes this can be resolved by having more than one PMO in the same business unit. There can be a separate PMO for IT activities and at the same time a corporate PMO for other activities. This can be effective if both PMOs work together, but more often than not they compete with one another, resulting in the failure of one or both PMOs.

Failing to Understand Cost of Paperwork

Project managers seem to understand the cost of paperwork while managing projects. But when assigned to a PMO, they often fall into the trap of believing that the PMO should be a paper-generating machine. The result can be an overwhelming increase in paperwork requirements for project teams to the point where PMO failure is inevitable.

The PMO should strive for paperless project management. As an example, some PMOs have converted their entire project management methodology to an intranet version which is entirely paperless. The entire methodology, together with the accompanying forms, guidelines, templates, and checklists, are on the company website rather than on paper. Project performance reporting can be done using a dashboard reporting system.

Failing to Understand Resource Capacity Planning

Most executives today do not know how much additional work they can undertake without overburdening the existing labor force. The PMO has the responsibility to provide this information to executives so that they can effectively establish a portfolio of projects based upon the availability of the necessary resources. Resource capacity planning could very well be the most important reason why executives support the existence of a PMO.

Failing to Understand That One Size Does Not Fit All

In a utopian environment, we would be able to create a single methodology or set of processes that could be applied to each and every project. In reality, this may be difficult to do because of the differences between projects. Forcing an organization to use improper processes can lead to a dislike for the PMO.

Failing to Understand That Processes Come Before Tools

Several years ago, before a particular company created a PMO, the company spent $600,000 for a licensing agreement for a certain project management software package. Later on, a PMO was established and given the charter to create a project management methodology for the company. It quickly became apparent that the software package that was selected was not a good fit for the methodology that was created.

Processes must come before tools. There are numerous software tools in the marketplace and many of the tools can even be customized to satisfy the needs of a specific project management methodology. When the wrong tools are purchased, blame seems to be placed directly upon the shoulders of the PMO regardless of where the decision was actually made to purchase the package.

Nonconsistent Use of Processes

An automotive industry supplier had three business units housed under one roof. Each business unit had its own way of managing projects. Problems occurred when some of the projects required that all of the divisions work together. Coordination of efforts became quite difficult for the project managers.

The president of the company established a PMO with the charter to create a methodology that could be used on a companywide basis. All three divisions supported the idea and assigned divisional resources to assist the PMO. One of the three divisions already had a methodology that the PMO believed could be used as a starting point. Eventually, a companywide methodology was created and was heavily based upon the information from one of the divisions.

When the final product was released and project teams were asked to use the new methodology, two of the three divisions argued against its use claiming that "it wasn't invented here." The PMO audited selected projects in all three divisions and found an inconsistent use of the processes; everyone was now blaming the PMO. Eventually, the president stepped in and mandated that the methodology and accompanying processes be used in all three divisions. Had the president not stepped in, it is entirely possible that the PMO may have been dissolved.

Failing to Establish PMO Metrics

PMOs are overhead rather than direct labor charges against various projects. As such, when economic conditions deteriorate, overhead activities are one of the first items to be looked at for possible cost reduction opportunities. This puts the PMO in danger of being dismantled.

PMOs must establish metrics that can show how the PMO adds value to the company and contributes to the company's bottom line. Typical metrics that the PMO might consider include:

- Percent of projects using/following the project management processes
- Ratio of the number of project managers to total project staff
- Better trained project managers
- Higher project success rates
- Improvements in customer satisfaction ratings
- Year-over-year throughput; doing more work each year with the same or fewer resources
- More efficient utilization of organizational resources
- The amount of decrease in the percent of projects at risk or in trouble
- Head count per project (staffing tolerance for projects)
- Finding ways to get faster closure
- Amount of reduction in the number of scope changes per project

Conclusion and Recommendation

The need for PMOs is quite apparent. But unless we fully understand the downside risks, the implementation process may not go as smoothly as we like. A PMO that does its job effectively will recommend changes for the better. People often tend to resist changes to their work habits. The PMO must address risk management issues for all decisions in order to prevent a PMO failure.

DISCUSSION QUESTIONS

The discussion questions are for classroom use to stimulate group thinking about PM 2.0. There are no right or wrong answers to most of the questions.

1. What bad things can happen if PMOs refuse to be networked together?
2. To whom should each of the PMOs discussed in the chapter report?
3. Should a PMO be allowed to manage projects? If so, what type of projects?
4. Should project managers be allowed to report permanently to a PMO while managing projects?
5. What skills should someone possess when assigned to a traditional PMO?
6. What skills should someone possess when assigned to a portfolio PMO?
7. What skills should someone possess when assigned to a strategic PMO?
8. What do you expect to see as the future of most PMOs?
9. Can PMOs succeed without executive support?
10. What can a PMO do to prevent its dismantlement during corporate downsizing efforts?

INDEX